Economic Models of Climate Cl

Economic Models of Climate Change

A Critique

Stephen J. DeCanio
Professor of Economics
University of California
Santa Barbara
USA

First published 2003 by
PALGRAVE MACMILLAN
175 Fifth Avenue, New York, N.Y. 10010 and Houndmills, Basingstoke, Hampshire RG21 6XS
Companies and representatives throughout the world

PALGRAVE MACMILLAN is the global academic imprint of the Palgrave Macmillan division of St. Martin's Press, LLC and of Palgrave Macmillan Ltd. Macmillan® is a registered trademark in the United States, United Kingdom and other countries. Palgrave is a registered trademark in the European Union and other countries.

ISBN 1–4039–6335–5 hardback
ISBN 1–4039–6336–3 paperback

This book is printed on paper suitable for recycling and made from fully managed and sustained forest sources.

A catalogue record for this book is available from the British Library.

Library of Congress Cataloging-in-Publication Data

DeCanio, Stephen J.
    Economic models of climate change : a critique / Stephen J. DeCanio.
        p.   cm.
    Includes bibliographical references and index.
    ISBN 1–4039–6335–5 – ISBN 1–4039–6336–3 (pbk.)
    1. Environmental policy – Economic aspects.   2. Climatic changes – Economic aspects.   I. Title.

GE170.D44 2003
363.738'74'011 – dc21

                                                                    2003040533

10   9   8   7   6   5   4   3   2   1
12   11   10   09   08   07   06   05   04   03

Printed and bound in Great Britain by
Antony Rowe Ltd, Chippenham and Eastbourne

For the future generations

Contents

List of Tables

List of Figures

Acknowledgments

Many colleagues and friends have contributed directly or indirectly to this book. As the notes and references throughout the text make clear, I owe a primary debt to those researchers who have communicated their insights through the scholarly literature. That literature has become so voluminous that I am sure I have missed many pertinent items; I apologize in advance for any such omissions.

A number of economists have contributed in a more personal way to my understanding of climate economics and modeling issues in general, particularly Gale Boyd, Richard Howarth, Florentin Krause, Paul Krugman, Richard Norgaard, Irene Peters, Alan Sanstad, and Jeffrey Williams. Members of the Economics Department at UCSB who have offered helpful advice are Henning Bohn, H.E. Frech III, Rajnish Mehra, and Jati Sengupta. Scholars outside the field of economics have provided much valuable input. Penelope Canan, Jeffrey Friedman, Jonathan Koomey, Kai Lee, Amory Lovins, Claudia Pahl-Wostl, Nancy Reichman, and Stephen Schneider have all widened my outlook. Friends outside of academia have also left their mark, including John Gliedman and Keith Witt. I relied on the advice of Diana Strazdes in selecting the cover art. I will never forget the inspiration and training I received from my professors at MIT over 30 years ago, especially Peter Temin and Franklin M. Fisher. I learned more than I was ever able to acknowledge from my deceased former colleagues M. Bruce Johnson and William N. Parker.

I want to offer special thanks to several of my former and current graduate students whose contributions have been invaluable. Keyvan Amir-Atefi provided criticisms and perspective throughout the project, and did an outstanding job of programming the evolutionary models described in Chapter 4. Yusuf Okan Kavuncu assisted in the programming for Chapter 3 and in the process significantly improved my understanding of overlapping-generations models. Andrea Lehman worked on some of the statistical analysis of Chapter 5, helping as much with material that was cut as with what remained. Catherine Dibble, William E. Watkins, Glenn Mitchell, and Ben Alamar showed me various aspects of the potential scope and power of computational methods in economics.

The book has a strong policy orientation. I first became involved in the economics of global environmental protection when I had the honor to serve as a Senior Staff Economist at the Council of Economic Advisers during 1986–87. Several people who were at the US Environmental Protection Agency during that time influenced the course of my career, including Steve Andersen, Eileen Claussen, and John Hoffman. In recent years, I have benefited greatly from the insight and patience of John A. "Skip" Laitner, Senior Economist for Technology Policy in the EPA Office of Atmospheric Programs, who was the Project Director of EPA Grant X-82830501-0 that provided support for a major portion of the project.

Some of the results from Chapter 2 have been presented at seminars and conferences held at the US Environmental Protection Agency, the UCSB Department of Geography, the Energy and Resources Group at the University of California at Berkeley, and the UCLA Computational Social Sciences Conference in Lake Arrowhead. Feedback from participants in those gatherings was greatly appreciated.

Acknowledgment is given to the World Resources Institute for permission to reproduce Figure 5.2, the International Project for Sustainable Energy Paths for Figure 5.3, and The Johns Hopkins University Press for most of Table 5.1 (Ascher, William. *Forecasting: an Appraisal for Policy-Makers and Planners*, pp. 130–1 © 1978. [Copyright Holder]. Reprinted with permission of The Johns Hopkins University Press.) Table 5.4 reprinted (with minor modifications and a change of units) from *Utilities Policy*, Vol. II, J.A. Laitner, S.J. DeCanio, J.G. Koomey, and A.H. Sanstad, "Room for Improvement: increasing the value of energy modeling for policy analysis," 87–94, © 2003, with permission from Elsevier.

The cover or dust jacket reproduces a painting by Martin Johnson Heade (American, 1819–1904), *Approaching Thunder Storm*, 1859, oil on canvas, 28 × 44 in (The Metropolitan Museum of Art, New York Gift of the Erving Wolf Foundation and Mr and Mrs Erving Wolf, 1975 [1975.160]. Photograph © 1992, The Metropolitan Museum of Art). Heade was associated with the second generation of Hudson River School landscape printers, artists who saw the condition of mankind reflected in the natural environment. *Approaching Thunder Storm* expresses foreboding in the time just prior to the American Civil War. Today our civilization is reflected not only in our view of nature, but is leaving an imprint on nature itself. Anthropogenic climate change looms as a global threat to human well-being. The dangers of climate change also span multiple generations. In Heade's painting, the older man on the shore is sitting by passively, but it is the youth in the boat who is at risk from the coming storm.

I want to express my appreciation to Amanda Watkins for her confidence in the book on behalf of Palgrave Macmillan, and to Philip Tye for expert copy-editing and assistance in production.

Finally, I owe the greatest debt to my family. My sons Jonathan, Samuel, and Aaron are worthy representatives of the first of the future generations. My loving wife, Annie Kirchner, is responsible for the domestic tranquility that is essential for the successful completion of a long project such as this. My parents, John and Alice DeCanio, also deserve a great deal of credit for starting me on the academic path. I only wish they could have lived long enough to hold copies of this book in their hands.

1
An Overview of the Issues

The responsibility of the present to the future is an abiding concern in human affairs. Much of our best effort is devoted to the upbringing of children, and they exercise a primary claim on our love and affection. The care, socialization, and education of the young is by far the largest "investment project" undertaken by any society. We work and build for the future, and strive to leave behind tangible legacies even beyond what we bequeath to our offspring. The onset of anthropogenic climate change challenges our link to the future in a very direct way – actions that are taken (or not taken) today will have an impact, possibly a decisive impact, on the condition of the natural world that will be inherited by those who follow us. While this is not a uniquely new "policy problem" (many of the great social and political issues have to do with matters affecting succeeding generations), climate change threatens human well-being across very long time spans in ways that are historically unprecedented.

Scientific research on the climate leaves no doubt that our actions (or inaction) are of vital significance for future generations. Each of the *Assessment Reports* compiled by the Intergovernmental Panel on Climate Change (IPCC)[1] has detailed the consequences of business as usual. Average global temperatures will rise, to be sure, but the magnitude of those increases (currently projected as between 1.4 and 5.8 °C [2.5–10.4 °F] from 1990 to 2100 [IPCC 2001a]) is not particularly indicative of the actual impact of climate change on human beings and other life on Earth. More specifically, unmitigated climate change will be a public health disaster of the first magnitude: people will die in killer heat waves, from the spread of tropical diseases like malaria as the range of the vectors carrying those diseases expands, from increased frequency and severity of floods, droughts, and possibly also tropical storms, from adverse effects on agriculture in some regions, and perhaps even from social disruption and conflict engendered by the climate change. These impacts will tend to fall disproportionately on the poorest segments of the world's population.

Both human and nonhuman systems will be affected. According to the most recent IPCC Synthesis Report, "[t]he stakes associated with pro-jected changes in climate are high" (italics in the original):

> Numerous Earth systems that sustain human societies are sensitive to climate and will be impacted by changes in climate (very high confidence). Impacts can be expected in ocean circulation; sea level; the water cycle; carbon and nutrient cycles; air quality; the produc-tivity and structure of natural ecosystems; the productivity of agri-cultural, grazing, and timber lands; and the geographic distribution, behavior, abundance, and survival of plant and animal species, including vectors and hosts of human disease. Changes in these systems in response to climate change, as well as direct effects of climate change on humans, would affect human welfare, positively and negatively. Human welfare would be impacted through changes in supplies of and demands for water, food, energy, and other tangi-ble goods that are derived from these systems; changes in opportu-nities for nonconsumptive uses of the environment for recreation and tourism; changes in non-use values of the environment such as cultural and preservation values; changes in incomes; changes in loss of property and lives from extreme climate phenomena; and changes in human health. Climate change impacts will affect the prospects for sustainable development in different parts of the world and may further widen existing inequalities. Impacts will vary in distribution across people, places, and times (very high confidence), raising important questions about equity.[2]   (IPCC 2001b, p. 238)

Even these dire effects do not tell the whole story.[3] Perhaps the great-est threat from climate change is the *risk* it poses for large-scale cata-strophic disruptions of Earth systems. Examples of such potential disasters include the shutting down of the oceanic "conveyor belt" that cycles warm water from the tropics to the North Atlantic off the coast of Europe, large reductions in the Greenland and West Antarctic ice sheets, accelerated global warming due to carbon cycle feedbacks in the terrestrial biosphere, and releases of terrestrial carbon from permafrost regions and methane from hydrates in coastal sediments. Again quoting the IPCC,

If these changes in Earth systems were to occur, their impacts would be widespread and sustained. For example, significant slowing of the oceanic thermohaline circulation would impact deep-water oxygen levels and carbon uptake by oceans and marine ecosystems, and would reduce warming over parts of Europe. Disintegration of the West Antarctic Ice Sheet or melting of the Greenland Ice Sheet could raise global sea level up to 3 m each over the next 1,000 years, submerge many islands, and inundate extensive coastal areas. Depending on the rate of ice loss, the rate and magnitude of sea-level rise could greatly exceed the capacity of human and natural systems to adapt without substantial impacts. Releases of terrestrial carbon from permafrost regions and methane from hydrates in coastal sediments, induced by warming, would further increase greenhouse gas concentrations in the atmosphere and amplify climate change.

(IPCC 2001b, p. 225, footnote omitted)

The probabilities of such devastating events are unknown, but are thought to be small. The risk is a combination of the *probability* and the *magnitude* of the impact, however. Facing the prospect of even very low probability events can be quite unpleasant, and people are generally willing to go to considerable lengths to mitigate such risks – as indicated by the widespread purchase of insurance of all types. In the case of climate risks, the affected Earth systems would not be amenable to crisis management, if they would respond to a sudden policy shift at all. Species cannot be brought back from extinction, and there is no way to restart the Atlantic conveyor belt or reassemble the West Antarctic ice sheet. Business as usual amounts to conducting a one-time, irreversible experiment of unknown outcome with the habitability of the entire planet.

Given the magnitude of the stakes, it is perhaps surprising that much of the debate about the climate has been cast in terms of *economics*. Economics does a relatively better job of calculating what might happen in the wake of marginal changes in policy or circumstances than it does in handling comparisons between radically different situations. Changes in the material standard of living may provide an adequate indicator of changes in human welfare if only small perturbations are being considered, and if the changes occur in the context of stable social, political, and cultural institutions. This requirement for "marginal" welfare analysis does not hold when considering climate change.

Preoccupation in the policy debate with economic arguments and calculations is even stranger in light of the fact that the climate problem is intrinsically one of intergenerational ethics. We do not look to technical economics for guidance regarding relations with our children and parents, or to specify the nature of the moral connections between the generations. Of course, some of these obligations are economic in nature – material support for the elderly and for children being the primary examples – but economic theory is not the source of our duties towards the young and the old. We shall see that the focus on economics in climate policy discussions is, in some sense, a conscious or unconscious attempt to avoid debating these difficult issues. After all, if a technocratic solution to one of the enduring dilemmas of the human condition could be found, why not embrace it? Economics offers to some the hope of finding such a purely technical answer, although we shall see that such hope is misplaced.

Nevertheless, it is undeniably the case that economic arguments, claims, and calculations have been the dominant influence on the public political debate on climate policy in the United States and around the world. Economic considerations were invoked by the Bush Administration in its repudiation of the Kyoto Protocol, and economic calculations informed the timid and defensive negotiating strategy of the Clinton Administration both before and after Kyoto. It is an open question whether the economic arguments were the cause or only an *ex post* justification of the decisions made by both administrations, but there is no doubt that economists have *claimed* that their calculations should dictate the proper course of action.

1.1 General equilibrium analysis

The current standard for economic analysis of large-scale policy issues such as climate is *general equilibrium analysis*. Unlike the "partial equilibrium analysis"[4] that is more familiar to the general public, general equilibrium analysis attempts to capture the essential features of the economic system as a whole. The intention is to trace the essential feedbacks between different sectors of the economy, and to create an analytical framework that reveals what might otherwise be the unintended consequences of policy actions.[5]

General equilibrium models purport to describe the key activities of production and consumption, as well as the market relationships that tie together the large array of goods, services, and factors of production that make up the economy. These models represent the production side of the economic system as a collection of profit-maximizing firms characterized by various technologies expressed as production functions.

The production functions are supposed to describe how combinations of inputs can be transformed into outputs. Firms seek to maximize their profits, defined as the difference between the revenues they obtain from selling their products and the cost of producing them. On the consumption side, general equilibrium models represent individuals as agents who atomistically seek satisfaction through the consumption of commodities. These agents' preferences are embodied in utility functions that exhibit certain features guaranteeing rationality of a particular sort. Utility is maximized subject to the requirement that an individual's spending cannot exceed a budget constraint that is a function of the individual's income and wealth.

This analytical framework emerged in the late nineteenth century, as economics began to employ mathematical methods to convey its concepts. The pioneers were Walras, Edgeworth, Jevons, and Marshall. It was unified and generalized in the twentieth century by the theorists Samuelson, Arrow, and Debreu, along with the others who formalized the neoclassical synthesis.[6] General equilibrium theory represents one of the pinnacles of achievement in economic thinking, and constitutes (along with closely related game theory) the foundation of modern economics. The general equilibrium model was conceived in the same spirit as the great syntheses of late nineteenth-century physics – the notion that it is possible to collapse the bewildering variety of real-world phenomena into the operation of a few powerful, abstract principles, thereby reducing the complexity of reality by embedding it in a compact mathematical structure. In this grand conception, empirical relationships are to be understood as manifestations of a few simple laws. Thus in physics, the combination of the universal conservation principles, Maxwell's equations of electrodynamics, and Newton's laws of motion (along with the inverse square gravitation law) provided a framework for working out accurate predictions and world-changing engineering applications in astronomy, communications, transportation, and manufacturing.

In economics, the unifying principles of general equilibrium theory are rationality and maximization. By specifying restrictions on utility functions to make them consistent with basic notions of rationality (such as transitivity or internal consistency and the existence of preference relationships between situations), and by deriving the consequences of maximizing behavior, the project of neoclassical economics was to reduce the description and understanding of economic phenomena to an elaboration of these basic principles. And indeed, an imposing intellectual edifice has been constructed. The goal is to represent the entire economic system by mathematical descriptions of the demand side of the economy (that is, the behavior of individuals as consumers of goods and services) and the supply side of the economy (that is, the behavior of firms and other producers), all linked together by equations expressing equilibrium conditions in the markets for the goods and services produced and consumed. To some, economics has been *defined* as the outcome of this program:

> Perhaps nothing is more readily distinctive about economics than the insistence on a unifying behavioral basis for explanations, in particular, a postulate of maximizing behavior. The need for such a theoretical basis is not controversial; to reject it is to reject economics. The reason such importance is placed on a theoretical basis is that without it, any outcome is admissible; propositions can therefore never be refuted. Economists insist that some events are *not possible*, in the same way that physicists insist that water will never run uphill. Other things constant, a lower price will never induce less consumption of any good; holding other productive inputs constant, marginal products eventually decline. There are to be no exceptions.
> (Silberberg 1990, p. 14; italics in the original)

We shall see in the subsequent chapters that even in strictly neoclassical terms, this kind of all-encompassing characterization of economic principles is far too rigid. More generally, it now is clear that the entire neoclassical project was overly ambitious. Rationality and maximization prove to be insufficient to characterize economic reality. Even within the boundaries of neoclassical economics, the hoped-for unifying principles are not enough to determine market outcomes. The sparse theoretical models of general equilibrium theory, despite their elegance, abstract from essential features of the actual social and economic system. Thus, the imagined kinship between economics and physics breaks down. While physics (and the physical sciences in general) have been successful in mathematical abstraction,[7] economics has not been. As we shall see, the simplifications of neoclassical economics strip away essential information about the system, not just the inessential accidentals. The consequences for climate policy have been severe.

The representations of consumers and firms that are the building blocks of the general equilibrium models employed in climate policy analysis lack the features that would make them realistic; or, going even farther, are so distant from the known behavior of actual individuals and businesses as to be implausible on their face. Furthermore, the mathematical structure built on the maximization principle, while beautifully elegant and interesting as an abstract exercise, turns out not to be sufficiently well-specified as to enable it to give the kind of policy advice – certainty about costs and benefits – that politicians desire. There are too many possibilities for multiple equilibria, unstable dynamics, and alternative distributional outcomes to pin down the economic system with enough precision to support policy recommendations based on neoclassical principles alone. Other assumptions, restrictions, or behavioral laws must be invoked to make the models well-behaved, and about these assumptions, restrictions, and behaviors there is no consensus. Nor is there any unambiguous empirical basis for choosing one particular set of assumptions or restrictions over another. The result is that the application of general equilibrium analysis to climate policy has produced a kind of specious precision, a situation in which the assumptions of the analysts masquerade as results that are solidly grounded in theory and the data. This leads to a tremendous amount of confusion and mischief, not least of which is the notion that although the physical science of the climate is plagued by uncertainties, it is possible to know with a high degree of certainty just what the *economic* consequences of alternative policy actions will be. This myth, more than any other, has created the policy paralysis and public confusion that so far have impeded constructive action (at least in the United States) to meet the climate challenge.

Instead of contributing its legitimate insights on the effects of various incentives, the interactions between different parts of the system, and the overriding importance of the distribution of wealth (more on this below), economics has been misused to obfuscate the climate debate. Economic models have been invoked to claim a knowledge of causes and consequences, of costs and benefits, and of the specifics of optimal policies, that are entirely beyond their grasp. Models routinely used in the policy arena involve forecasts and projections extending decades into the future, but in reality no economic forecasting technique has any hope of embodying accurate information about circumstances that far ahead. Models are used to compare policy alternatives, but the fundamental principles of economics make those models incapable of carrying out the requisite comparisons. Models are claimed to represent economic and social reality, despite the fact that it is known that they omit, ignore, or mischaracterize vast segments of that reality. Models are used to make strong statements about which policies should or

should not be undertaken, even though it is known that at their foundations, the mathematical properties of the models preclude drawing
welfare conclusions. The subsequent chapters of this book will discuss
the basis of all of these assertions.

1.2 Equity and efficiency

It should not be thought that neoclassical economic theory (and empirical work based on it) has nothing to offer. The absence of a "theory of
everything" does not mean that no scientific lessons have been learned.
An apt analogy is that of Paul Krugman, who compares the present state
of economics not to physics, but to medicine *circa* 1900.[8] Even though
medicine at that time could not claim an understanding of health and
disease based on the "microfoundations" of molecular biochemistry
(nor can it today in most cases), medical practice nevertheless was based
on a number of hard-won insights. The same is true of economics today.
We know, among other things, the benefits of decentralizing many economic decisions, the importance of aligning individuals' incentives and
policy goals, and the key role played by technological change in raising
standards of living.

To illustrate the kind of economic insight that has largely been
ignored in the climate debate, consider the relationship between the
concepts of equity and efficiency. Equity and efficiency are the twin
poles of neoclassical theory. Equity has to do with the distribution of
wealth and income, while efficiency is concerned with getting the most
out of any particular set of resources. Most formal economic modeling
having to do with climate policy has focused on efficiency issues, even
though it is disputes over equity that have plagued the international
negotiations and have made it impossible so far to arrive at a domestic
policy consensus.

Although there are circumstances in which equity and efficiency concerns may properly be separated, the climate debate is not one of them.
As will be shown in more detail in the following chapters, the potential allocations of various kinds of "rights" relevant to climate policy
are so important that they affect all significant matters of price and
allocative efficiency. To pretend otherwise amounts to an implicit
commitment to a particular set of choices about equity. The distribution of rights across generations, and within different groups of people
presently alive (rich or poor in the United States, for example, or North
or South in the world) is so important that prices, interest rates,
incomes, and welfare all depend on the way the rights are allocated.

This will be illustrated in subsequent chapters through a series of very simple general equilibrium models that show the connections. Ironically, although the large integrated assessment general equilibrium models[9] that essentially ignore equity have been the most influential economic contributions to the debate so far, it is easy even in very simple general equilibrium models to bring equity issues to the fore-front. The later chapters will show why the allocations of climate rights across time, space, and income class determine the most salient features of climate policy and its consequences.

The development of welfare analysis in economics has been a long struggle to establish the limits of what economics could say regarding social arrangements. The culmination of this quest is represented by the fundamental theorems of welfare economics. As stated succinctly in Mas-Colell et al. (1995), these are:

> *The First Fundamental Welfare Theorem.* If every relevant good is traded in a market at publicly known prices (i.e., if there is a complete set of markets), and if households and firms act perfectly competitively (i.e., as price takers), then the market outcome is Pareto optimal. That is, when markets are complete, *any competitive equilibrium is necessarily Pareto optimal.*
>
> *The Second Fundamental Welfare Theorem.* If household preferences and firm production sets are convex, there is a complete set of markets with publicly known prices, and every agent acts as a price taker, then *any Pareto optimal outcome can be achieved as a competitive equilibrium if appropriate lump-sum transfers of wealth are arranged.*
>
> (p. 308, italics in the original)

Mas-Colell and his co-authors go on to explain that

> [t]he first welfare theorem . . . is, in a sense, the formal expression of Adam Smith's claim about the "invisible hand" of the market. The second welfare theorem goes even further. It states that under the same set of assumptions as the first welfare theorem plus convexity conditions, all Pareto optimal outcomes can in principle be implemented through the market mechanism. That is, a public authority who wishes to implement a particular Pareto optimal outcome (reflecting, say, some political consensus on proper distributional goals) may always do so by appropriately redistributing wealth and then "letting the market work".   (1995, p. 308)[10]

There are technical subtleties lurking within these definitions: the "convexity" required by the second theorem rules out increasing returns to scale such as are known to exist at least at the level of firms and perhaps industries. "Price-taking" or competitive behavior by firms and households rules out unrestrained self-seeking behavior by monopolies or oligopolies. "Publicly known prices" means that everybody has full information about the prices for all the goods and services that are being transacted in all markets all the time. Obviously, these stringent assumptions are not likely to hold in the real world. At the same time, economists have long held that they provide a kind of standard against which the actual performance of the economy can be measured, and as such have formed the basis of antitrust legislation, truth in advertising regulations, and prohibitions against insider trading or other forms of deceptive or collusive economic behavior.

As important as potential deviations from the competitive ideal might be, they are not going to be the focus of attention in this book. Instead, the consequences of alternative distributions of wealth of different kinds will be worked out in simple, stylized models that accept the assumptions underlying the two welfare theorems.[11] The importance of "complete markets" is that the environmental circumstances affecting people's well-being have to be subject to exchanges – market transactions in other words – and that all the people affected be able to participate in those transactions somehow. In ordinary economic terms, this means that there must be "property rights" in all the material things that matter to people. To avoid seeming to be too narrow, in what follows the "property" part will be dropped and reference will be made only to "rights," as in "climate rights," "emissions rights," and so forth. The second welfare theorem becomes important when it is realized that, even if all the conditions for market equilibrium and Pareto optimum are realized, the social outcome that is actually observed will depend on the *allocation of rights* of all types.

Property rights originate with the government, because it is the government that defines what kinds of actions are lawful, what kinds of exchanges are permitted, and what kinds of contracts are enforceable. The process by which the State makes these decisions is of course vital, but whether a government is democratic or authoritarian, welfare/reformist or socialist, constitutional monarchy or majoritarian republic, the sovereignty of the State constitutes the foundation of the definition of rights. These definitions are not unchangeable. As recently as the mid-nineteenth century, slavery was legally recognized in the United States. The slave laborer did not own the right to the proceeds of his or her labor, and was not free to change employers if the conditions of work for a particular slave owner became too onerous. Other labor systems are characterized by other configurations of rights. In serfdom, the serfs are not free to leave the land to which they are assigned, but they are free to obtain whatever price may prevail in the market for their produce. In free labor markets, workers are able to switch jobs and are entitled to retain the market value of their labor (before taxes). They may also have the right to due process protections against arbitrary dismissal, entitlements to unemployment or health insurance, and so on. Obviously, there is a wide spectrum of rights that can be assigned to the different parties participating in labor markets, and the assignment of these rights is determined by the law – that is, by the State.

The abolition of slavery may be the most dramatic example of how property rights can change, but it is not the only one. The limited-liability corporation, an innovation necessary to enable the agglomeration of large amounts of capital needed for industrial-scale productive enterprises, was a legal innovation. Today, the courts are struggling with defining new kinds of property rights – in genetic information or in the data stream that makes up a recorded musical performance, for example. Nor are these the only kind of new rights that are economically important. A great deal of social policy swirls around "entitlements" of various sorts – to a particular level of state-funded pension benefits, to certain medical services, etc. Not all of these rights can be traded in markets. Social entitlements typically are inalienably attached to individuals. Yet the practical significance of those entitlements depends on interpretation and enforcement of laws. Ultimately, it is the State that makes this determination.[12]

The reason this matters for climate policy is because the future outlines of the economy are going to be determined, to a very large degree, by the kinds of rights – in climate stability, emissions levels, or fossil fuel use – that ultimately will be policy-determined. The situation until now has been one in which users of fossil fuels have been free to dispose of the waste products of the combustion of those fuels (mainly CO_2) for free. No one owned the atmosphere; there were no regulations on fossil-fuel burning, and there was no price associated with increasing greenhouse gas loadings on the atmosphere. This allocation of "climate rights" was appropriate in the preindustrial and early industrial world, when energy demands were relatively low and human activity did not have much of an impact on the atmosphere as a whole. The free disposal of fossil fuel combustion wastes contributed to the Industrial

Revolution by enabling the solar energy stored in fossil fuels to be converted cheaply into useful work.

That situation no longer prevails. Today, the human impact on the climate (and the natural world more generally) has become massive and measurable. The consequences are severe, both in terms of likely future damages and in terms of the risk of catastrophic surprises. The environmental impact of human activities is so profound that the current geological era can be called the "Anthropocene" (IPCC 2001a, p. 784, citing Crutzen and Stoermer 2000). If and when governments begin to address the consequences, and assign various kinds of environmental or climate rights to people (including future potential victims of climate change), the result will be a change in the allocation of wealth. This reallocation will significantly affect the outcome of market processes. Prices, interest rates, and incomes all will be influenced. The insights of general equilibrium theory (and of the two fundamental welfare theorems) give an indication of how the economic system will reflect the new allocation.

Yet these fundamental alterations in wealth holdings are systematically downplayed by the practices of current integrated assessment modeling.

- Models based on "representative agents" rule out the possible consequences of allocations to different kinds of people. In the real world, individuals vary in their preferences and their endowments of other types of wealth (natural abilities, current holdings of different kinds of property, etc.). Policies adopted or not adopted will change relative endowments of environmentally related forms of wealth;
- Market outcomes based on current definitions of property rights are treated as the standard for welfare comparison, even though it is known that welfare depends on the pattern of allocation of *all* rights, including those presently undefined that give rise to externalities;
- The pattern of allocations of rights affects the characteristics of market equilibria, including whether those equilibria are unique and stable. Without a comprehensive treatment of all the rights that make up individuals' endowments, analysis of the equilibria will be incomplete and is likely to be misleading.

The dream of neoclassical economics was to establish a "theory of value," a framework in which observable quantities and prices could be connected to people's tastes and desires and to the technologies of production. In the case of systems having a unique equilibrium, this goal can be approached, with price ratios equal to the ratios of marginal utilities and the first and second welfare theorems holding. But multiple equilibria wreck the project. Alternative sets of equilibrium prices can satisfy the marginal conditions, but with completely different distributions of income (even with the same set of endowments). Hence, the connection between observable market quantities and the "fundamentals" of human preferences is severed. The different equilibrium configurations, all of which would be Pareto optimal, correspond to very different social orders.

The problem is just as bad or worse with general equilibrium models incorporating the time dimension. Such models may exhibit a multiplicity of steady-state equilibrium solutions, and in addition, there can be a continuum of equilibrium price paths approaching the steady states. Hence, very little can be deduced from the evolution of prices over time regarding the well-being of the people. The real social choice problem is between equilibrium configurations, not about marginal changes within a particular system, and economics has little to offer in the way of guidance.

In the chapters that follow these points will be developed at length. It will be shown that the rights allocation problem applies both at any particular time and over time. Furthermore, a realistic portrayal of production leads to other sources of multiplicity and ambiguity in model outcomes. Examples will be given of simple models exhibiting counterintuitive properties, depending on how the rights to different goods are distributed.

1.3 Outline of the book

Chapter 2, "The Representation of Consumers' Preferences and Market Demand," is devoted to how individuals' preferences or tastes are expressed in climate policy models. The chapter has two main parts. The first is devoted to what might be called the "outside critique" of the neoclassical utility function representation. In this section, the kinds of arguments that have been raised against the utility-function-based mathematical versions of "economic man" are reviewed. The second part of the chapter takes all the standard neoclassical assumptions as given, then develops simple general equilibrium exchange models that exhibit properties that call into question the way conventional climate policy analysis is carried out. The necessary aggregation of individual demand functions into market demand functions cannot be guaranteed to yield a well-behaved system. Specific examples of multiple equilibria and unstable dynamics are worked out in detail. It is shown how some of the untested features of standard energy/economic models that are usually taken for granted are in fact crucial *assumptions* that determine the results of the modeling.

Chapter 3, "The Treatment of Time," extends the ideas of Chapter 2 to models in which time is treated explicitly. It is shown that the long time periods over which climate policy must be analyzed create the very conditions under which the multiplicity of equilibria and instability of dynamics are likely to arise. This is developed both in Arrow–Debreu and overlapping-generations frameworks. It is shown that, so long as no particular time is selected as a preferred vantage point, the equilibria may differentially favor any of the generations that now exist or will come into being in the future. The question of whether there is a preferred time vantage point (such as, for example, the present) is an ethical question that cannot entirely be settled within economics. The *absence* of a preferred time vantage point is akin to the physical principle of relativity theory, that there is no preferred coordinate system and that physical laws should be independent of the particular coordinate system in which their equations are expressed. This chapter also clarifies the debate over whether (and how) future costs and benefits should be discounted, and does so in a unified framework that incorporates previous approaches to this controversial issue.

Chapter 4, "The Representation of Production," shifts the discussion to the supply side of the economy. The current state of knowledge about the behavior of firms is reviewed, focusing on the question of whether firms can validly be treated as entities that maximize profit subject to their production functions. The modern theory of the firm does not support this characterization, nor does the evidence on the relative efficiencies of firms. The chapter goes on to suggest that evolutionary models of industrial dynamics hold more promise for providing a sound basis for analyzing production, and gives examples of how such evolutionary models could be set up (with an emphasis on computability), as well as the kinds of results that can (and cannot) be derived from such models.

Chapter 5, "The Forecasting Performance of Energy-Economic Models," takes up a related question: Even if the theoretical bases for the consumption and production components of climate economic models are suspect, might they nevertheless have enough predictive power to be useful in the formulation of policy? The chapter takes advantage of the fact that models of the economy that emphasize energy production and consumption have been in use since the 1970s, when the first oil price shocks drew attention to the significance of the energy sector. The performance of these models can be evaluated over a considerable number of years (approximately three decades). It is shown that no matter what the forecasting interval, the models have almost no predictive power. In addition, models that have been used to forecast the cost and impact of a range of environmental and other regulatory measures do not do well in prediction either.

Chapter 6, "Principles for the Future," is a recapitulation of the main results and brings together the policy recommendations that have been presented in each of the preceding chapters. It offers a summary of how economic knowledge might more fruitfully be brought to bear on the climate problem. The conclusion is that economists would gain in credibility, and their recommendations would be more valuable to governments and citizens grappling with the complexities of the climate issue, if economics were more modest in its claims.

2
The Representation of Consumers' Preferences and Market Demand

2.1 Introduction

The general equilibrium models used for climate policy analysis are stylized representations of the activities of the millions of individuals and organizations that constitute the economy. The models themselves are made up of systems of equations that represent production and consumer demand, and spell out the market conditions that determine the prices and quantities of goods bought and sold. In some cases, key features of the models are determined outside the interactions they describe, or "exogenously." For example, technological progress (which can be measured as the increase in output that can be obtained from given inputs as time goes on) is often specified as a constant percentage rate of change independent of other variables in the model. Technical progress can also take the form of entirely new products or services. The conditions of general equilibrium determine how the structural and behavioral equations are to be solved to yield the prices and quantities that emerge within the economy. The meaning of "general equilibrium" is that all markets clear in the sense that the plans and intentions of consumers and producers are fulfilled.

This chapter will focus on the way *consumers' preferences* are handled in such models. A critique can be made at two levels. The entire concept of treating individuals as self-contained, rational utility maximizers, with their preferences taken as given (that is, determined outside the model), is a departure from realism. Similarly, the underlying definitions of property rights, and the existence of the markets that enable the individuals' preferences to be made manifest, assume a great deal about the constitution of society. The rejection of these kinds of economic abstractions might be called the "outside" critique, because it entails standing

16

outside the economics framework and asking whether the conventional assumptions made by economists make sense in the first place. In addition, however, there is an "inside" critique – points that fall entirely within the conventional formalism of economics and are generally acknowledged by economists themselves. The inside critique reveals that even if all the standard abstractions of neoclassical theory are accepted, the mathematical structure that results contains many pitfalls and ambiguities that are usually not taken into account in conventional climate policy analysis. The consequence is that the "results" of the conventional analysis are dependent to a much greater degree than is usually recognized on a set of assumptions for which there is little or no scientific evidence.

2.2 Elements of the "outside" critique

While the way consumer behavior is treated within economics has considerable intuitive appeal, the intellectual structure supporting it is quite elaborate. For example, the notion of "rationality" in economics requires that individuals have well-defined preferences over all different combinations of goods and services, and that these preferences are "transitive" (that is, if A > B and B > C, then A > C where A, B, and C represent different consumption bundles) (Mas-Colell et al. 1995). These requirements are by no means innocuous – we all know of situations in which people simply cannot make up their minds, or make choices that are apparently inconsistent (such as when people change their minds about an action or a purchase), or in which having more choices is actually worse than having a clear guideline for action.[1]

In addition, the translation of the "axioms of rationality" into scientific propositions about relative prices and responses to price changes is predicated on the existence of commodities that are priced and tradable. If something important (such as climate stability) is neither traded nor priced, there is no way of using real-world information about consumer behavior to compare marginal shifts in expenditure on this commodity with spending on other goods. In such cases, to employ economic techniques requires some method for imputing quantifiable values. Economists often employ proxies such as the "value of time" or the "statistical value of a premature death avoided" to approximate the value of environmental goods. In other cases, survey information (the "contingent valuation" technique) is used to assign dollar values to things that matter to people but for which markets do not exist.

Although the "outside" critique calls the assumptions embodied in these methods into question, it also extends more broadly. Even restricting the discussion to economic *categories* is limiting. Essential elements of human behavior as it pertains to climate include the widest range of considerations of culture, motivation, and social organization (Jochem et al. 2000; see also Jacobs 1994). What are some of the directions in which these criticisms have been developed?

2.2.1 The exogeneity of preferences

Even before discussing the rationality of individual preferences, a prior question is the origin of the preferences themselves. It might seem obvious that the beliefs, values, and tastes of human beings are not formed independent of the social context. While it is clear that we all have basic needs arising from our physical nature – requirements for nourishment, shelter, and contact with other persons – there can be no doubt that a large segment of our mental makeup is socially constructed from our upbringing, experiences, and culture (Brekke and Howarth 2000, 2002). It is an unjustified (and unjustifiable) analytical simplification to treat people's preferences as determined outside the social landscape. Yet neoclassical economics makes just this leap; it makes no attempt to analyze or understand how or why some material goods are learned to be desirable while others are devalued.

A related and perhaps even more fundamental point is that no system of thought based on analyzing the happiness derived from material goods can adequately address the ultimately philosophical question of what constitutes "the good." It is undeniable that many of the most important things that affect well-being are "commodities" only under the most encompassing of definitions. Family and community relationships, the welfare of one's children, environmental quality, personal security, and good health are "commodities" only if the meaning of that term is stretched almost beyond recognition. And, of course, happiness is not obtained through the acquisition of commodities alone.[2] Furthermore, the concept of "the good" transcends happiness. Moral and ethical principles can (and sometimes must) supercede considerations of personal satisfaction. Heroic deeds, such as those performed by the New York police and firefighters, or the airline passengers who resisted the hijackers on September 11, are not measured by a utility-maximization calculus. Climate policy extends into the realm of ethics too, because the consequences of decisions made by people now alive will affect others not yet born. Thus, no analysis of climate policy can be complete if it is based solely on the preferences of those now living.

The fact that "the good" involves more than material well-being does not, of course, diminish the value of economics as a means of gaining insight into human affairs. Production, trade, and consumption are essential components of life, and the pursuit of happiness or enlightenment is difficult without a base of material security. A society's economy might be compared to the plumbing in a house – the plumbing is not the main determinant of the well-being of the house's occupants, but it is important that the plumbing function well. Even so, the "Integrated Assessment" of climate change (and the design of policies to address climate change) goes beyond the workaday operation of the economy because climate stability – and global environmental protection generally – involves the whole of humans' physical surroundings and the fate of the entire biosphere. Of course, economic theory can be expanded to cover all human activities: "leisure" can be treated as a commodity to be consumed; clean air, climate stability, and biodiversity likewise. But the more the scope of economic analysis is expanded to include such things, the less tenable is the presumption that preferences can simply be taken as given.

The treatment of tastes as exogenous is particularly noninnocuous with respect to climate change. The consequences of climate change may occur to people distant from us spatially and temporally. Whether or not such impacts "matter" to us is a question of ethics and values that is very far removed from the creature-based cravings for food and shelter that are perhaps least dependent on culture. Hence, to initiate a discussion of "the economics of climate change" starting from the presumption that individual tastes and preferences are given from outside the system distorts the nature of the problem.

In the context of policy analysis, the assumption of the immutability and exogeneity of tastes imparts a peculiar form of conservative bias to the exercise. If tastes are given, there is no legitimate room for education or political persuasion. Thus, the notion that the people in a democratic society might be convinced that they should change their behavior or institutions in response to an environmental threat is ruled out. Even if the educated elite were to grasp the technically complex arguments and information necessary to see an impending climate problem, the elite would have no role, within the confines of "economic analysis," for imparting their superior insight to the masses. Of course, no one in a well-functioning democracy would operate as if this were the case. Discussion, debate, and argument are essential features of a healthy polity. Tolerance means an honest acknowledgment of differences and recognition of the rights of others, not an indifference to the path of social development or the fate of one's fellow human beings. Yet strict adherence to the immutability of preferences would deny the reality (and effectiveness) of the interpersonal communication that is ubiquitous in human societies.

2.2.2 Markets require property rights

Suppose, however, that we were willing to accept preferences as given. Conventional economic analysis then entails a working out of the terms of exchange (prices) and allocations of productive resources given those preferences. Markets are the social mechanisms by which this is accomplished. Markets are defined by the exchange transactions that take place between individuals, and a precondition for such exchanges is the existence of well-defined property rights in the commodities that are being exchanged. Property rights and the associated rules for their enforcement are nothing other than a way of specifying the spheres of control of the agents in the economy. My property right in my home enables me to exclude others from its use; my right to exchange my labor for income is a way of ensuring that I am fairly compensated for my efforts (provided there are a number of employers willing to compete for my services, and that I have the freedom to choose between job offers).

The liberal tradition places a high social value on market transactions because of the welfare implications of their being *voluntary*. Voluntary transactions are guaranteed to *improve the well-being of both parties*, because if they did not, they would not take place.[3] Nevertheless, this ideal outcome should not be assumed to govern every eventuality. In particular, there may be no "property rights" associated with some of the things that impinge on a person's well-being. This absence of complete property rights results in "externalities." Usually these externalities are treated as an exception to the general cases encompassed by the economic model, but in fact they are endemic.[4] Where global climate stability is at stake, no system of property rights now exists that enables individuals to express their preferences for one kind of climate regime versus another; nor would it be a simple matter to set up a system that would enable the market exchange paradigm to achieve anything like a desirable outcome. Nevertheless, the definition and enforcement of appropriate property rights are a social and political problem that must be solved prior to the successful functioning of an economic system based on market transactions and exchange.

Numerous policy proposals are being advanced to create property rights suitable for climate protection. The Kyoto Protocol, by specifying national greenhouse gas emission limits, represents a step in this direction. The "property right" to emit CO_2 and the other controlled greenhouse gases for countries adhering to the Protocol is defined in terms of a particular percentage of those countries' 1990 levels of emissions. The countries participating in the Kyoto system will be able to conduct a limited amount of trade of their emissions rights. Although there is no apparent movement within the United States to adhere to Kyoto, several proposals to reduce greenhouse gas emissions by defining new property rights are in play. One example is the "Sky Trust" (Barnes and Pomerance 2000, Barnes 2001). A national emissions limit would be specified with emissions permits assigned to a permanent trust. Each year, dividends (arising out of revenues from the permits) would be distributed to the citizens of the US on an equal per capita basis.[5]

A similar plan for distribution of greenhouse gas emissions permits on an equal *global* per capita basis has been advanced by EcoEquity (Athanasiou and Baer 2001). An outline of their proposal was recently published in the policy forum of *Science* (Baer et al. 2000). Other suggestions involving creation of new rights include placing a cap on the prices of emissions permits issued by governments (Kopp et al. 1997), or the McKibbin–Wilcoxen proposal that would create, in each country, two kinds of assets – an emission permit required by fossil fuel industries to supply a unit of carbon annually and an emission endowment giving the owner an emission permit every year forever. Under the McKibbin–Wilcoxen plan, the price of the annual permits over the first few years would be fixed by international negotiation (thereby controlling potential short-run costs), while the price of the perpetual endowment would reflect expectations of future permit prices (to be determined periodically by renegotiation in light of scientific and technical information), much as a stock certificate reflects expected future dividends. A significant portion of the initial allocation of the endowment could be given to the fossil fuel industries to enlist their political support for the proposal (McKibbin 2000). Of course, the status quo also represents an implicit assignment of rights: as things now stand, anyone has the right to use the atmosphere for disposal of CO_2 and other greenhouse gasses at zero charge.

It will be shown subsequently that the creation and assignment of these kinds of rights will have a profound impact on the shape of economic activity over time. In an interdependent (general equilibrium) economic system, the pattern of rights ownership affects prices, incomes, and allocations of all goods and services. For the moment, it is sufficient to observe that market transactions cannot guarantee individuals' well-being when property rights in vital commodities have not been defined. Consider the case of conventional cost–benefit analysis (CBA).[6] In CBA, the costs of environmental protection (measured in terms of reductions in marketed outputs or increases in the costs of production of a given level of output) are compared to some kind of monetized measure of benefits. The implicit justification for this approach is in the standard economic representation of the equilibrium of the consumer. In equilibrium, the ratios of marginal utility to price for each good are equal. The equality of these ratios across all commodities means that, at prevailing prices, the consumer gains the same additional or marginal utility from expenditure of a dollar of income on any of the commodities. Alternatively, the consumer is indifferent to subtracting a dollar of expenditure from one of the commodities and spending it on another. The argument is that if this "indifference condition" did not hold, the consumer could increase his utility by rearranging expenditures.

But how can such reasoning be applied to commodities (climate stability, air pollution levels, or biodiversity) for which no markets and no property rights exist? There is no social determination of the "prices" at which these "commodities" might be transacted, because they are not exchanged at all. The levels of risk from climate change or loss of biodiversity that people bear are purely a consequence of other activities undertaken in response to other incentives (such as the prices for inputs and outputs that do prevail in real markets). Hence, to conduct a CBA, prices have to be *assumed* or *imputed* for the environmental goods. A variety of techniques are employed for this purpose.

For example, wage differentials in jobs requiring similar qualifications but having different levels of risk can be assumed to represent the disutility of risk in general. The wage differences therefore represent "compensating variation" for the differing levels of risk associated with the different jobs. Similarly, the price of safety devices (smoke detectors, automobile air bags) might be taken as a measure of how much people are willing to pay to avert certain kinds of risk. There are several problems in applying this approach to valuing the risks of climate change, however. The value for the "price of risk" obtained in different markets varies by as much as an order of magnitude (Viscusi 1993). Part of the reason has to do with selection; a willingness on the part of some individuals to work in a risky industry such as Alaska fishing does not mean that the wage premium offered for that work would be sufficient to entice most people to take on the risk. In addition, if avoidance of risk is an ordinary good, then willingness to bear risk should decrease with wealth, compounding the problem of extracting risk preferences from wage data. It is also the case that people have different attitudes towards different types of risk. It is much more unpleasant to bear a risk imposed without one's consent than to undertake a risk voluntarily. Most people would hate to have a nuclear power plant sited in their neighborhood, but (at least some of) those same people are willing to pay large sums of money to risk life and limb on the ski slopes. Climate change falls into the category of an imposed risk as opposed to one freely undertaken. Also, it is not at all clear that people make informed estimates of the risks of various activities. For many years, people resisted using safety belts in automobiles because they feared being trapped in a burning wreck, when in fact the risk of fatality or serious injury from being thrown from the car (or from smashing into interior surfaces) was many times greater. In the case of climate change, even specialist experts are not sure of the nature and magnitude of the risks; it is impossible for an ordinary citizen to know them. Finally, nothing in the theory of "risk pricing" explains why people undertake deadly activities such as smoking.[7] It seems quite plausible that these behaviors are conditioned by a combination of misperceptions of risk and "social" effects such as peer group pressure or status-seeking.[8]

There is another method used to impute prices to nontransacted environmental commodities. This method is known as contingent valuation or CV. In CV analysis, people are asked what they would be willing to pay for (or how much would be required to compensate them for) the use or loss of an environmental benefit. A variety of survey techniques can be used, including the "referendum"-style question in which people are asked to give a yes or no answer to the question of whether they would pay a named amount to prevent an environmental loss such as a particular climate change scenario. There are two fundamental limitations to this approach as applied to climate issues. The first is that only people alive today can be surveyed, and hence the response data pertains only to members of the current generation. The second is that the climate problem is so complex, with regard to possible policy approaches but especially with regard to the consequences of climate change, that it is unrealistic to expect ordinary citizens to be able to give informed responses to any survey no matter how well designed. Expert opinion on the magnitude, timing, and risks of climate change varies. Only a mechanical faith that "democracy" or "public opinion" can somehow miraculously aggregate disparate, partial, and frequently conflicting information into a coherent and reasonable policy can justify reliance on survey information to serve as a guide for policy. This is not to say that public opinion and beliefs should not be relevant, only that the responses of the relatively uninformed public to a policy challenge cannot be guaranteed to produce a good outcome. This is an instance of the phenomenon in political theory known as "public ignorance" that has strong implications for what can and cannot be expected from democratic political systems (Friedman 1997, DeCanio 2000a).

In addition to these fundamental limitations, there are very serious technical problems that limit the applicability of CV information to climate policy analysis. These technical problems include the presence of "protest votes" (in which respondents refuse to answer the question) and the "embeddedness problem" (that the willingness to pay (WTP) for a set of environmental values appears to be no larger than the WTP for one of the parts).[9] In one famous example, Desvousges et al. (1993) found a WTP to avoid killing birds that was similar for saving 2000, 20,000, or 200,000 birds. The WTP methodology must exclude feelings of altruism or public-spiritedness from individuals' responses. Otherwise, WTP would imply that income should be redistributed in favor of people who care about each other. These and other conceptual and empirical difficulties led Diamond and Hausman to say:

> In short, we think that the evidence supports the conclusion that to date, contingent valuation surveys do not measure the preferences they attempt to measure. Moreover, we present reasons for thinking that changes in survey methods are not likely to change this conclusion. Viewed alternatively as opinion polls on possible government actions, we think that these surveys do not have much information to contribute to informed policy-making. Thus, we conclude that reliance on contingent valuation surveys in either damage assessments or in government decision making is basically misguided.   (1994, p. 46)

Most significant for climate issues (and independent of the controversy over the technical issues in survey design) is the problem that CV questions are predicated on the *existing definitions and distribution of property rights*. A poor farmer in a developing country might be willing to pay very little to implement a regulatory policy to reduce greenhouse gas emissions worldwide, but his view would be different if the carbon-reduction policy entailed creation of "atmospheric rights" to greenhouse gas emissions, and the distribution of those rights on an equal per capita basis globally (as called for in the EcoEquity plan, for example). "Willingness to pay" under existing property rights can be seen almost entirely as an added burden, while a policy of distributing atmospheric rights would involve creation of a valuable new form of wealth.

The only way to obtain market information on individuals' preferences regarding climate change or other important environmental goods is through the creation and assignment of property rights in those goods. Even the existence of well-functioning markets is not enough to guarantee good social outcomes, however. The key advantage of market transactions is that they are voluntary and hence putatively beneficial to both parties to the transaction. However, the very notion of what is voluntary cannot be separated from the definition of the property rights and the assignment of their ownership. People can voluntarily exchange what belongs to them, but the category of "belonging" is socially and legally determined. Thus, a laissez-faire State might be one that defines and enforces private ownership of titles to land, shares of stock in corporations, and bonds, while a Welfare State may define entitlements to minimum incomes or particular forms of medical care. In the laissez-faire State there might be no noncoercive way to obtain medical care if an individual lacks the wealth to pay for it; in general, voluntary transactions are possible only if both parties to the transaction want something legitimately owned by the other party. A person who, for example, lacked sufficient wealth to obtain some desired medical treatment could only obtain the treatment by "coercively" violating the health care providers' rights. In the Welfare State, on the other hand, entitlements equivalent to property rights enable transactions to take place that otherwise would not, and "coercion" takes the form of State-sanctioned redistributive measures necessary to support the entitlements. The difference is not that one type of State is coercive and the other is not; rather, the difference is in which property rights are defined and enforced (Friedman 1990).

The voluntary transactions of the market system, while improving the well-being of the transactors, ultimately have indeterminate effects on other members of the society. The price changes that accompany technological progress inevitably have *distributional* effects. Consider an example. Before electronic computers, there was a significant market for mechanical calculating machines. The Economics Department at MIT in the late 1960s had a room filled with unused electromechanical calculators, capable only of addition, subtraction, multiplication, and division, neatly arrayed in rows, forlornly awaiting gangs of graduate students who would never come again to employ them for calculating regression statistics. The machines themselves were ingenious contraptions of keys, wheels, and carriages. Within a few years after the onset of programmable electronic computers, they were gone, assigned either to landfills or to museum exhibits. The human and physical capital that had been employed in their manufacture was devalued in the "creative destruction" (Schumpeter's memorable phrase) that accompanied the rise of electronic computers. This type of effect is known in economics as a "pecuniary externality," and is the inevitable accompaniment of technological change. In this example, the holders of stock in the makers of the calculators were losers, while those who had invested in IBM were gainers (other things equal).

How do we know that "technological progress" is welfare-improving? There appears to be no a priori way of concluding that it must be. It is possible to imagine a type of technological change that would come to dominate the market and yet would reduce people's well-being. For example, surveillance technology could lead to complete real-time monitoring of the movements and activities of all citizens, and such technologies might be adopted in response to social demands (to deter running red lights, to enable parents to know where their children are at all times, to speed paramedics to heart-attack victims, to prevent terrorism, etc.), but at the same time the universal adoption of those technologies could lead to a lowering of welfare (through loss of privacy and police-state controls). Purely private-sector examples can also be imagined – it seems likely that the diffusion of antibacterial home cleansing products only speeds up the evolution of resistant strains of pathogens. The fact that technological progress *has been* welfare-improving over humanity's historical experience is an *empirical observation*, not a logical necessity.

Finally, it should always be kept in mind that voluntary market exchanges are not the only ways in which economic transactions can be mediated. Relying on its power of taxation and monopoly of legitimate force, the State can appropriate wealth for its own purposes (purposes which may or may not be supported by the populace).[10] The most important transactions within families take place according to rules of reciprocity and mutual affection that have little to do with market mechanisms. A great many interactions within organizations are shaped by relative power relations, or by the requirements of bureaucratic procedures. Market transactions are an important category of social activity, but can hardly be construed to be the archetype of human interaction.

Economic modeling practice largely ignores these varieties of human interpersonal relations and focuses on market or market-like exchanges.[11] In doing so, however, more is required than the elevation of some parts of the human personality (and certain social institutions) to center stage while ignoring all the rest. In addition to the radical autonomy and well-defined property rights required for a market-oriented world view, economists also assume that people adhere to the economic version of rationality (see above). This kind of narrow definition of rationality is both a strength and weakness of conventional economic theory. It is a strength because it leads to testable proposi- tions, a requirement of any genuine science.[12] It may also be appropri- ate as a normative stance (applied to some individual choice problems, intended to support decision-making). It is a weakness because it forces economics into a posture in which very large segments of human behav- ior necessarily lie outside consideration.[13]

It should also be noted that preferences in the standard neoclassical model are entirely "atomistic," that is, an individual's utility is a func- tion only of his own consumption of goods and services, and does not depend in any way on the consumption of others. There are at least two reasons why this extreme version of individualism is not valid. First, we know that the well-being of others does influence our own happiness. Radical altruism is not required; even economics recognizes that the household is more plausible as a decision-making unit than the indi- vidual, because of the close personal ties between household members. But even beyond the household, it is implausible that we are not affected by the situation of others with whom we come in contact. A given material standard of living will produce different levels of utility if one is surrounded by armies of beggars and destitute homeless people as opposed to encountering other individuals of comfortable means. It does not matter if the adverse impact of the poverty-stricken is treated as an "externality" in the modeling of preferences; the result is to make each person's utility dependent to some degree on the general social level.

Of equal force is the observation that people derive utility from so- called *positional goods*, goods whose values depend on the individual's possession of them relative to other members of society. One version of this is the notion that *relative consumption* enters the utility function, not just the absolute level of consumption. It is well known that the existence of relative consumption effects or positional goods destroys the Pareto optimality of the general equilibrium (Howarth 1996, Frank 1985a, see also Frank 1985b). Yet the presence of these effects is surely ubiquitous. If any economist doubts this, he or she should consider the preoccupation of Economics Departments with their relative academic rankings.

This chapter will not revisit the issues surrounding economic rationality, or the other elements of the "outside" critique. Instead, the limitations imposed by the economic description of human activity will be taken as working hypotheses. What is not commonly understood is that *even within the strictly economic domain, the range of possible social outcomes is much greater (and more complex) than the conventional neoclassical computable general equilibrium (CGE) models used for integrated assessment suggest.* If one accepts all of the assumptions regarding individual rationality, the availability of information, the exogeneity of tastes and preferences, and the kinds of transactions that are allowed to take place, the outcome in models of general equilibrium is *remarkably open-ended.* That is, the standard economic assumptions, even if they apply perfectly to the individual agents making up the economy, are not sufficient to specify the properties of the economy as a whole. The implications of the assumptions for *individual* behavior do not, in general, carry over to a comparable set of implications for *aggregate* behavior.[14] This means that general equilibrium economic models as presently constituted cannot provide unambiguous climate policy guidance. To force "answers" out of such models necessitates invoking a number of nonscientific and untested *assumptions.*

2.3 The "inside" critique

In order to make rational consumers part of applied general equilibrium models, the standard practice is to describe the consumers' behavior in terms of a "utility function."[15] The consumer's utility function is a theoretical fiction that provides a starting point for the derivation of observable relationships. Positing utility functions for the consumers bypasses all the questions of how preferences are formed, whether individuals behave rationally, whether tastes can be described fully in terms of well-defined commodities, and all the other metacriticisms of economic methodology touched on previously. In what follows, all the neoclassical assumptions regarding the existence and consistency of preferences are taken as given.

The utility functions of consumers can never be observed directly (and in principle *cannot* be observed, because the utility function is specified only up to a monotonic transformation); only the consequences of utility maximization – the individual demand functions – can possibly correspond to observable phenomena. (The demand functions relate quantities purchased and consumed to relative prices and the wealth or income of the consumer.) Equilibrium prevails when aggregate demand (defined as the sum of the individuals' demands) equals aggregate supply for each commodity. The fact that market-clearing equilibrium is defined in terms of the *aggregate* demand and supply functions for the commodities is the Achilles' heel of applied general equilibrium models. Economic theory shows that aggregate demand functions cannot be guaranteed to have the same properties as individual demand functions. In fact, aggregate demand functions can have almost any shape at all. One implication of this key theoretical result is that it is not proper to lump consumers together as a "representative agent" whose demand behavior is the same as that of an individual. Another implication is that aggregate demand functions based on perfectly rational individual behavior can lead to multiple equilibria and unstable dynamics, either of which would be sufficient to call into question the kinds of uses to which the models are put in climate policy analysis.

2.3.1 Multiple equilibria

These theoretical results have been well-known since the 1970s.[16] They are the culmination of a century of development of economic ideas. The results were derived by the most distinguished economic theorists and published in the most accessible journals. There is no professional "dissensus" on the validity of the proofs. One might think that in the normal course of scientific progress, these insights would be incorporated into the state-of-the-art models used for policy analysis. Yet the reality is that, for the most part, these results have essentially been *ignored* by policy-oriented economists. There has been no concerted effort to determine whether the conditions under which a general equilibrium representation of the economy is well-behaved (that is, whether it has a unique and stable equilibrium) prevail in actuality. Instead, the economists whose CGE integrated assessment models are most influential in climate policy circles have simply *assumed* conditions that guarantee that their models will be well-behaved. Those assumptions are usually made implicitly and without discussion, and have become acceptable in the peer-reviewed literature as the "standard" of practice. This is not the same thing, unfortunately, as their being valid.[17]

Just how serious is this problem? A good way to illustrate what is at stake is to examine some small general equilibrium models that display the problematic behavior. By exploring the factors that make for trouble (like multiple equilibria, dynamic instability, etc.), one can see what would have to be tested for or established empirically if the project of using general equilibrium models for climate policy analysis were to be solidly grounded. It will also be possible to illustrate how different prior visualizations of how environmental goods enter people's preferences may account for the apparently irreconcilable perspectives of economists and environmentalists. The behavior of the examples will shed light on the policy debate, because as we shall see, the conventional integrated assessment models either *rule out by assumption the possibilities of greatest concern to environmentalists*, or are based on *empirical research that could not possibly reveal the underlying conditions that make the troublesome outcomes appear.*

For simplicity, we will consider first exchange economies in which there is no production.[18] Each of the consumers begins with an endowment of one or more of the commodities in the economy, and each behaves as a price taker. In other words, there is no strategic behavior, no formation of alliances or cartels, and no problems of information or uncertainty. The examples will be ones in which there are only a few consumers and commodities. Consumers' preferences will be represented by utility functions of the "constant elasticity of substitution" (CES) type. This is a standard functional form that enables the degree of substitutability of the goods to be characterized by a single parameter in the utility function.

The elasticity of substitution between commodities indicates how large a change in the optimal consumption mix would result from a change in the relative prices faced by the consumer. Commodities are "substitutes" if a small change in their relative prices leads to a large change in the relative quantities consumed; the commodities are "complements" if a large change in relative prices results in only a small shift in the relative consumption of the two goods. Intuitively, goods are complements if the enjoyment of each is highly dependent on possession of the other; they are substitutes if either can be enjoyed without regard to the other. In standard economic textbooks (and in the commonsense lay understanding), substitutes are goods that fulfill the same function: home appliances of different colors are substitutes; public transport plus a bicycle is a substitute for a car; and food purchased at the supermarket and prepared at home is a substitute for restaurant meals. Complementarity of goods also has a commonsense interpretation. Left and right shoes are complements. In a less trivial example, for most parents their own consumption and that of their children are highly complementary – it is difficult to imagine parents who could enjoy a high standard of living if their children were impoverished.

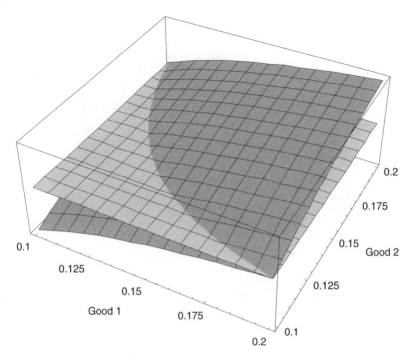

Figure 2.1(a) Utility surface for two goods with high elasticity of substitution

Similarly, the enjoyment of automobiles is dependent on the existence of a smoothly functioning street and highway system; if roads are perpetually clogged by traffic jams and construction projects, little pleasure can be derived from driving.

A simplified illustration of complementarity and substitutability is given in Figures 2.1(a) and (b), giving a picture of the consumer's utility function in the case of two goods. The light gray horizontal plane in both figures is drawn for a particular level of utility. The intersection of the utility surfaces (dark gray) with the utility level would trace out the familiar two-dimensional indifference curves for each of the two utility functions.

For our purposes, one key empirical issue is the elasticity of substitution between environmental goods of various types and for the ordinary goods and services that are transacted in markets. The conventional economist's view is that environmental goods are substitutes with respect to marketed goods. This is implicit in CBA, where

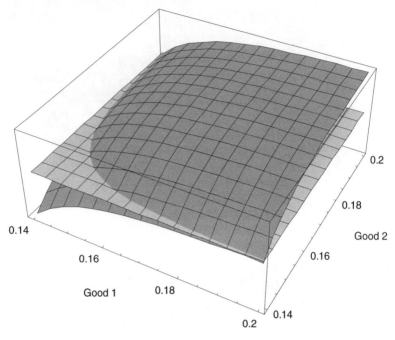

Figure 2.1(b) Utility surface for two goods with low elasticity of substitution

policies are evaluated by comparing the (imputed) value of environmental benefits to the cost of environmental protection policies. For policies involving large reallocations (such as would be required for climate stabilization, for example) this criterion requires essentially perfect substitutability between the environmental goods and marketed goods (as well as the presumption that the imputed price of the environmental goods is exactly right).[19] In the environmentalist view, on the other hand, environmental quality is highly *complementary* to the goods obtained through the market – even a very large increase in the consumption of marketed goods could not compensate for climate instability or the massive loss of biodiversity.

Of equal or greater concern for climate policy is the degree to which goods can be substituted *over time*. This is measured by the intertemporal elasticity of substitution, a parameter about which there has developed a considerable econometric literature. Chapter 3 will contain a full discussion of the treatment of time in climate models, so a further discussion of the intertemporal elasticity of substitution will be deferred until then.

To move beyond graphical and verbal descriptions, it is necessary to set up a simple general equilibrium model. The starting point of such a model is the mathematical description of the individual consumers' utility functions. In the first simple model that will be discussed,[20] the utility functions will take the form

$$u_i = \sum_{j=1}^{n} \frac{a_{ij} x_{ij}^{b_i}}{b_i} \tag{2.1}$$

where u_i is the utility of the ith individual, x_{ij} is the consumption of good j by individual i, and the parameters a_{ij} and b_i describe the preferences of individual i. The a_{ij} may be thought of as the weights that individual i attaches to each commodity's consumption, while the b_i parameter captures the degree of substitutability between commodities in individual i's utility. This utility function is of the constant elasticity of substitution type because the parameter b_i is common to all commodities for individual i. The index n denotes the number of commodities, and in the models that will be considered here this will also indicate the number of agents (i.e., there will be the same number of agents and goods).[21]

The degree of substitutability could in principle differ between goods for a given consumer. The assumption that the same b_i is common to all goods is a simplification for expositional and mathematical convenience.[22] Formally, the elasticity of substitution between any two goods is defined as the percentage change in the ratio of the optimal quantities of the two goods consumed, in response to a percentage change in their relative prices, holding the level of utility constant. That is, the elasticity of substitution η_{jk} between goods j and k (for individual i) is defined as

$$\eta_{jk} = \frac{\partial \log\left(x_{ij}^* / x_{ik}^*\right)}{\partial \log(p_j / p_k)} \tag{2.2}$$

where x_{ij}^* and x_{ik}^* are the utility-maximizing values of goods j and k consumed at prices p_j and p_k. It is easy to show that for the utility functions specified in equation (2.1), this elasticity of substitution between any two goods for the ith consumer is given by

$$\eta_{jk} = \frac{1}{1 - b_i} \tag{2.3}$$

For values of b_i ranging from 0 to $-\infty$, the elasticity of substitution ranges from 1 to zero; for values of b_i greater than zero but less than 1, the elasticity of substitution ranges from 1 to $+\infty$.[23] Goods are complements for low values of the elasticity of substitution; they are substitutes for relatively high values.[24]

The initial conditions for the exchange economy are completed by specifying the endowments of each consumer. (Remember, there is no production, so all the goods originate as endowments held by the consumers.) The ith consumer's endowment of the jth commodity is given by ω_{ij}. The prices of the commodities are p_j ($j = 1, \ldots, n$). *Consumer behavior* is specified by the condition that each consumer maximizes utility, subject to the budget constraint determined by the consumer's endowments. That is, each consumer solves the problem,

$$\max_{x_{i1}, x_{i2}, \ldots, x_{in}}(u_i) \quad \text{subject to} \quad \sum_{j=1}^{n} p_j x_{ij} - \sum_{j=1}^{n} p_j \omega_{ij} \leq 0 \tag{2.4}$$

Solving the system of first-order conditions obtained from the consumers' maximization problems results in individuals' demand functions in which the demands for each commodity are a function of the prices and the endowments. These individual demand functions are denoted by starred values, with $x_{ij}^*(p_1, p_2, \ldots, p_n; \omega_{i1}, \omega_{i2}, \ldots, \omega_{in}, b_i)$ being the demand of individual i for good j. Solving the first-order conditions yields utility maxima for the individuals because the utility functions are all convex to the origin.

Market equilibrium is then determined by the condition that the "market excess demand" for each commodity be equal to zero. That is, the total amount of each commodity demanded by the consumers (obtained by summing their individual demands) is equal to the total amount of that commodity available in the economy (obtained by summing the individual consumers' endowments). This condition amounts to the requirement that

$$f^j(p_1, p_2, \ldots, p_n; \forall a_{ij}, \omega_{ij}, b_i)$$
$$= \sum_{i=1}^{n} x_{ij}^*(p_1, p_2, \ldots, p_n; \omega_{i1}, \omega_{i2}, \ldots, \omega_{in}, b_i) - \sum_{i=1}^{n} \omega_{ij} = 0 \tag{2.5}$$

for all n goods j. The excess demand functions f^j are functions of all prices as well as all the individual endowments and the parameters of the utility functions. It is known from basic price theory that these functions are homogeneous of degree zero in the prices, that is, only relative prices matter. (This must be so because the set of feasible consumption bundles for any individual consumer cannot change if all prices are multiplied by a scalar constant, and the market demand function is made up of the sum of the individual consumer demand functions.) Homogeneity of degree zero implies that there are only $n - 1$ independent equations like (2.5). One of the goods can be taken as the numeraire and its price set, so that the $n - 1$ independent equations can be solved for the remaining prices. Alternatively, a price normalization rule (such as requiring that the prices sum to one) can be imposed without altering the substantive results of the model.

The setup of the model economy is completed by specification of the numerical values of the parameters. First, consider the very simple case of two consumers and two goods, as described by Kehoe (1998). In Kehoe's example, the parameter values are $a_{11} = a_{22} = 1024$, $a_{12} = a_{21} = 1$, $\omega_{11} = \omega_{22} = 60$, $\omega_{12} = \omega_{21} = 5$, and $b_1 = b_2 = -4$. With good 2 taken as numeraire, this economy has three equilibria. The three values of p_1 are 0.127, 1.0, and 7.856. To see the source of the multiplicity of equilibria in this simple example, consider how various combinations of prices and parameter values can satisfy the market-clearing condition of equation (2.5). Suppose that $a_{11} = a_{22}$, and the other parameter values are as in Kehoe's example. Figure 2.2 shows a three-dimensional plot of the excess demand for good 1, as a function of p_1, for different values of a_{11}.

In Figure 2.2, the light gray plane represents the market equilibrium condition, $f^1 = 0$. The equilibria corresponding to different values of a_{11} are the points at which the dark gray surface (the market excess demand function) intersects the $f^1 = 0$ plane. For low values of a_{11}, the excess demand surface cuts the $f^1 = 0$ plane only once for each a_{11}, and there is only a single equilibrium (at $p_1 = 1$). However, for values of $a_{11} > 41.7$ (approximately), there are multiple equilibria as the excess demand function cuts the $f^1 = 0$ plane three times for each value of a_{11}.

This example of Kehoe's is instructive, but it does not convey the extent of the possibilities for multiple equilibria inherent in simple general equilibrium models. Consider a similar but somewhat expanded example with five consumers and five goods. Again, all five consumers have symmetric utility functions, with $b_i = -4$ ($i = 1, \ldots, 5$), $a_{ii} = 1024$ ($i = 1, \ldots, 5$), and $a_{ij} = 1$ ($i \neq j$, $\forall i, j$). The initial endowments of the commodities are given by $\omega_{ii} = 60$ ($i = 1, \ldots, 5$) and $\omega_{ij} = 1$ ($i \neq j$, $\forall i, j$). Prices are normalized to sum to one.[25] The specified values of b_i correspond to elasticities of substitution of 0.2 between all pairs of goods for all consumers. In this simple case, the economy has *at least 31 distinct equilibria*. These are displayed in Table 2.1.[26] Because the endowments are symmetrical and the prices add to one, GDP will be the same in all

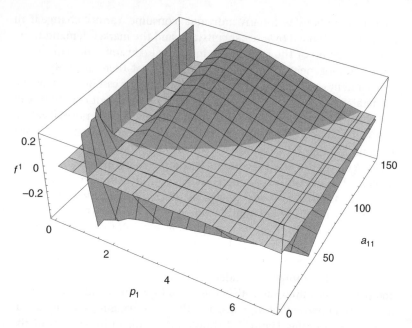

Figure 2.2   Excess demand for good 1 as a function of p_1 and a_{11}

the equilibria in this economy.[27] The utility rankings in this case correspond to the individual income levels y_i for each of the five agents. (For
each agent i, $y_i = \Sigma_j p_j x_{ij}$.) As will be seen below, this correspondence
between utility rankings and individuals' money incomes does not hold
in general.

In any event, it is clear that there is no way to rank the equilibria
unambiguously in terms of their social desirability. Equilibria 6–10 show
the greatest inequality, and are most preferred by the single (different)
"winners" and least preferred by the four "losers" in each case. Any one
of these five equilibria would seem quite pleasant to the person who
happened to be at the top of the income distribution, but would be
miserable (relatively speaking) for everyone else. Observe that every
kind of income distribution is possible: one high-income agent and four
low-income agents (equilibria 6–10); two high-income agents and three
low-income agents (equilibria 22–31); three relatively high-income
agents and two relatively low-income agents (equilibria 11–20); four
relatively high-income agents and one relatively low-income agent
(equilibria 1–5), and the completely egalitarian distribution (equilibrium

Table 2.1 Equilibrium prices, incomes, and Gini index of inequality, symmetrical 5 × 5 economy

Equilibrium	Prices					Incomes					Utility rankings (8 = best)					Gini
	p_1	p_2	p_3	p_4	p_5	y_1	y_2	y_3	y_4	y_5	u_1	u_2	u_3	u_4	u_5	
1	0.067	0.233	0.233	0.233	0.233	5.0	14.8	14.8	14.8	14.8	3	5	5	5	5	0.122
2	0.233	0.067	0.233	0.233	0.233	14.8	5.0	14.8	14.8	14.8	5	3	5	5	5	0.122
3	0.233	0.233	0.067	0.233	0.233	14.8	14.8	5.0	14.8	14.8	5	5	3	5	5	0.122
4	0.233	0.233	0.233	0.067	0.233	14.8	14.8	14.8	5.0	14.8	5	5	5	3	5	0.122
5	0.233	0.233	0.233	0.233	0.067	14.8	14.8	14.8	14.8	5.0	5	5	5	5	3	0.122
6	0.973	0.007	0.007	0.007	0.007	58.4	1.4	1.4	1.4	1.4	8	1	1	1	1	0.713
7	0.007	0.973	0.007	0.007	0.007	1.4	58.4	1.4	1.4	1.4	1	8	1	1	1	0.713
8	0.007	0.007	0.973	0.007	0.007	1.4	1.4	58.4	1.4	1.4	1	1	8	1	1	0.713
9	0.007	0.007	0.007	0.973	0.007	1.4	1.4	1.4	58.4	1.4	1	1	1	8	1	0.713
10	0.007	0.007	0.007	0.007	0.973	1.4	1.4	1.4	1.4	58.4	1	1	1	1	8	0.713
11	0.047	0.047	0.302	0.302	0.302	3.8	3.8	18.8	18.8	18.8	3	3	6	6	6	0.281
12	0.047	0.302	0.047	0.302	0.302	3.8	18.8	3.8	18.8	18.8	3	6	3	6	6	0.281
13	0.047	0.302	0.302	0.047	0.302	3.8	18.8	18.8	3.8	18.8	3	6	6	3	6	0.281
14	0.047	0.302	0.302	0.302	0.047	3.8	18.8	18.8	18.8	3.8	3	6	6	6	3	0.281
15	0.302	0.047	0.047	0.302	0.302	18.8	3.8	3.8	18.8	18.8	6	3	3	6	6	0.281
16	0.302	0.047	0.302	0.047	0.302	18.8	3.8	18.8	3.8	18.8	6	3	6	3	6	0.281

Table 2.1   *Continued*

Equilibrium	Prices					Incomes					Utility rankings (8 = best)					Gini
	p_1	p_2	p_3	p_4	p_5	y_1	y_2	y_3	y_4	y_5	u_1	u_2	u_3	u_4	u_5	
17	0.302	0.047	0.302	0.302	0.047	18.8	3.8	18.8	18.8	3.8	6	3	6	6	3	0.281
18	0.302	0.302	0.047	0.047	0.302	18.8	18.8	3.8	3.8	18.8	6	6	3	3	6	0.281
19	0.302	0.302	0.047	0.302	0.047	18.8	18.8	3.8	18.8	3.8	6	6	3	6	3	0.281
20	0.302	0.302	0.302	0.047	0.047	18.8	18.8	18.8	3.8	3.8	6	6	6	3	3	0.281
21	0.200	0.200	0.200	0.200	0.200	12.8	12.8	12.8	12.8	12.8	4	4	4	4	4	0
22	0.464	0.464	0.024	0.024	0.024	28.4	28.4	2.4	2.4	2.4	7	7	2	2	2	0.488
23	0.464	0.024	0.464	0.024	0.024	28.4	2.4	28.4	2.4	2.4	7	2	7	2	2	0.488
24	0.464	0.024	0.024	0.464	0.024	28.4	2.4	2.4	28.4	2.4	7	2	2	7	2	0.488
25	0.464	0.024	0.024	0.024	0.464	28.4	2.4	2.4	2.4	28.4	7	2	2	2	7	0.488
26	0.024	0.464	0.464	0.024	0.024	2.4	28.4	28.4	2.4	2.4	2	7	7	2	2	0.488
27	0.024	0.464	0.024	0.464	0.024	2.4	28.4	2.4	28.4	2.4	2	7	2	7	2	0.488
28	0.024	0.464	0.024	0.024	0.464	2.4	28.4	2.4	2.4	28.4	2	7	2	2	7	0.488
29	0.024	0.024	0.464	0.464	0.024	2.4	2.4	28.4	28.4	2.4	2	2	7	7	2	0.488
30	0.024	0.024	0.464	0.024	0.464	2.4	2.4	28.4	2.4	28.4	2	2	7	2	7	0.488
31	0.024	0.024	0.024	0.464	0.464	2.4	2.4	2.4	28.4	28.4	2	2	2	7	7	0.488

Source: See text.

21). The Gini coefficient of income inequality varies dramatically across the equilibria. The Gini index ranges from 0 for equilibrium 21 (the perfectly egalitarian income distribution) to the highly inegalitarian income distribution of equilibria 6 through 10 (Gini of 0.713).[28] Equilibria 22 through 31 have an intermediate degree of income inequality, with a Gini index of 0.488. For comparison, from the end of World War II through the early 1990s, the Gini coefficient for inequality of household income in the United States has ranged from a low of 0.348 in 1968 to a high of 0.429 in 1992 (Weinberg 1996).

Each of these equilibria is a Pareto optimum: no consumer could be made better off without decreasing the utility of at least one other consumer. The fact that the first-order conditions hold at each of the equilibrium points guarantees this result.[29] The multiplicity of Pareto optimal equilibrium points does not mean, however, that the consumers would be indifferent between such points, nor does it imply which equilibrium would result from decentralized economic decision-making. If the members of society could somehow "vote" on which equilibrium they would prefer, any of the 10 equilibria (11–20) could command a majority. This majority would not necessarily be stable politically, however. Every majority of three relatively well-off agents corresponds to one of the equilibria 11–20. Any one of these majorities of three would be vulnerable to reshuffling, as one of the disadvantaged minority could offer part of his income as "side payments" to two members of the majority as an inducement for them to form a new majority coalition including the previously disadvantaged individual.[30] This kind of reshuffling is hypothetical, of course; there is no way that a society could even know about the existence and properties of all the equilibria, let alone devise a procedure for choosing among them.

The different equilibria arise because the individuals have preferences that *differ*, they have endowment patterns that *differ*, and the goods are *different* (i.e., they are not good substitutes for each other). The parameters of the utility functions are such that the individuals' demand functions exhibit strong wealth effects. For an individual well-endowed with a particular good, an increase in the price of that good can increase the individual's wealth sufficiently to offset the "substitution effect" that would tend to cause the individual to substitute other goods for the one whose price has increased. Such wealth effects are not unknown even for nonenvironmental goods – perhaps the most common illustration is the backward-bending labor supply curve (the case in which a sufficient increase in the wage induces the worker to work less and consume more leisure).

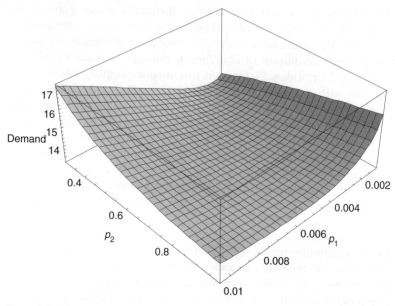

Figure 2.3 Demand surface for agent 1 and good 1, p_3, p_4 and p_5 at equilibrium 7

An illustration of the operation of the wealth effect in the first agent's demand for good 1 is given in Figure 2.3. This figure shows part of the demand surface for good 1 by agent 1 as a function of the first two prices, p_1 and p_2, with the other prices held constant at their values in equilibrium 7. Only part of the demand surface is shown for a limited range of the two prices. For very low values of p_1, the substitution effect (of a change in p_1) dominates and the demand function is downward sloping as p_1 increases, but the wealth effect comes to predominate for larger values of p_1 and the demand function increases with p_1. Figure 2.3 does not tell the whole story of the demand by agent 1 for good 1, of course. It is impossible to visualize geometrically the full shapes of the individual demand functions or the even more complex market demand functions in six dimensions (the five price dimensions plus the quantity demanded).

The parameters of the utility functions in these examples are such as to create strong wealth effects. However, the possibility of multiple equilibria in general equilibrium models is not just an artificially conjured special case. As Kehoe puts it,

Non-uniqueness of equilibrium does not seem so pathological a situation as to warrant unqualified used of the simple comparative statics method when dealing with general equilibrium models. . . . [F]or most of these models no method now exists to determine whether the equilibrium found . . . is unique. To make matters worse, it appears that non-uniqueness of equilibria is an even more common situation in applied models that allow for such distortions as taxes, price rigidities, and unemployment than it is in the simple model. . . .   (1985, p. 145)

There is very little "lore" about the conditions under which the multiplicity of equilibria arises in practical applications.[31] It certainly is not possible to examine the expenditure shares of the different goods in the economy and conclude from them that the ones with low shares in the total value of output are "less important" than those with high shares. In the model of Table 2.1, the expenditure shares of the different goods are the same as their relative prices because the endowments are completely symmetrical and the total quantity of each good is 64 units. But in equilibrium 6, for example, the expenditure share of good 1 is 97.3 percent, while the shares of goods 2 through 5 are 0.7 percent each. Good 1 makes up the lion's share of GDP, while goods 2, 3, 4, and 5 each make up less than 1 percent of GDP. However, this is entirely a consequence of the fact that the economy happens to be in equilibrium 6; in equilibrium 7, good 2 has the overwhelming expenditure share while the shares of the other goods are small. The expenditure shares of the different goods vary across each of the 31 equilibria. Yet the utility functions and endowments are *entirely symmetrical*, so it would be sensible to maintain that all the goods are equally "important." If multiple equilibria exist, there is no way to infer the relative importance of the goods to individuals from the expenditure shares of the goods.

Some work has been done to explore the question of multiplicity in applied models (see, for example, Kehoe and Whalley 1985), but even if the computational burden were not formidable, checking existing models for multiple equilibria would not really address the problem. The reason is that the issue of multiple equilibria is often settled in the *design* of the models without reference to any underlying economic reality, and therefore is not an issue that can be explored by examination of finished models. We will return to this issue below.

Now consider the same type of 5×5 model economy, but with one minor change. All the utility functions are the same as in the first example, and all the endowments are the same with the exception that agent 1's endowment of the first good is 70 rather than 60. This economy has at least 25 equilibria, as displayed in Table 2.2.

Table 2.2 is similar to Table 2.1 in that the agents' utility rankings correspond to their income levels. Unlike in the previous model, however, the GDP is not the same in all the equilibria. Because agent 1 has a slightly larger endowment, the GDP values computed for the entire economy follow the same ranking as the utility ranking for agent 1. This surely does not correspond to any kind of "social" ranking of the equilibria however; it is evident from Table 2.2 that four out of the five agents making up the economy would prefer any equilibrium (with the exception of 12–15) to equilibrium 11, the one having the highest GDP. (These relative rankings are independent of the price normalization rule, again because the relative prices are not dependent on the normalization rule chosen.) As in Table 2.1, if the agents could vote on which equilibrium the economy should reach, a majority could be put together that would support one of the set from 16 through 19, although the "divide the pie" type of political instability discussed before makes it impossible to determine a particular majority-preferred equilibrium within that set. The GDP of these equilibria are in fact the third-lowest of the possibilities, because agent 1 fares poorly. The voting rules matter as well. If a supermajority of four votes were required for the society to pick an equilibrium, it would settle on equilibrium 1, which has a higher GDP than any of the set 16–19 that would be reached by simple majority rule.

A third example 5×5 model differs from the first two in that the consumers have different elasticities of substitution, as well as different endowments. As in the first two examples, there are five consumers and five goods. The weights in the consumers' utility functions are the same as before, but now $b_1 = -3$, $b_2 = -3$, $b_3 = -4$, $b_4 = -5$, and $b_5 = -5$. The endowments are given by $\omega_{11} = \omega_{55} = 70$, $\omega_{22} = \omega_{33} = \omega_{44} = 60$, and $\omega_{ij} = 1$ ($i \neq j$, $\forall i, j$). This third example also exhibits multiple equilibria. The consumers' utility functions are not symmetric, so the equilibria will not fall into symmetrical groups as did many of the equilibria in Tables 2.1 and 2.2. At the same time, because the numerical search for equilibria is sensitive to the starting point (and possibly also to the numerical search method), there is no guarantee that all the equilibria are displayed in Table 2.3.[32] The table contains the values for prices, incomes, and utility rankings for five equilibria that were found, however. As in the case of Table 2.2, the ranking of equilibria by GDP tells us nothing about the social desirability of the equilibria. The GDP

Table 2.2 Equilibrium prices, incomes, and utility rankings, 5 × 5 economy, asymmetrical endowments

Equilibrium	Prices					Incomes					Utility rankings (8 = best for agent 1, 13 = best for agents 2–5)					GDP
	p_1	p_2	p_3	p_4	p_5	y_1	y_2	y_3	y_4	y_5	u_1	u_2	u_3	u_4	u_5	
1	0.017	0.246	0.246	0.246	0.246	2.2	15.5	15.5	15.5	15.5	4	10	10	10	10	64.1693
2	0.553	0.149	0.099	0.099	0.099	39.1	9.8	6.9	6.9	6.9	6	8	6	6	6	69.5290
3	0.553	0.099	0.149	0.099	0.099	39.1	6.9	9.8	6.9	6.9	6	6	8	6	6	69.5290
4	0.553	0.099	0.099	0.149	0.099	39.1	6.9	6.9	9.8	6.9	6	6	6	8	6	69.5290
5	0.553	0.099	0.099	0.099	0.149	39.1	6.9	6.9	6.9	9.8	6	6	6	6	8	69.5290
6	0.594	0.216	0.063	0.063	0.063	42.0	13.7	4.7	4.7	4.7	7	9	5	5	5	69.9405
7	0.594	0.063	0.216	0.063	0.063	42.0	4.7	13.7	4.7	4.7	7	5	9	5	5	69.9405
8	0.594	0.063	0.063	0.216	0.063	42.0	4.7	4.7	13.7	4.7	7	5	5	9	5	69.9405
9	0.594	0.063	0.063	0.063	0.216	42.0	4.7	4.7	4.7	13.7	7	5	5	5	9	69.9405
10	0.549	0.113	0.113	0.113	0.113	38.9	7.7	7.7	7.7	7.7	5	7	7	7	7	69.4866
11	0.937	0.016	0.016	0.016	0.016	65.7	1.9	1.9	1.9	1.9	8	2	2	2	2	73.3732
12	0.003	0.978	0.007	0.007	0.007	1.2	58.7	1.4	1.4	1.4	1	13	1	1	1	64.0287

Table 2.2 Continued

Equilibrium	Prices					Incomes					Utility rankings (8 = best for agent 1, 13 = best for agents 2–5)					GDP
	p_1	p_2	p_3	p_4	p_5	y_1	y_2	y_3	y_4	y_5	u_1	u_2	u_3	u_4	u_5	
13	*0.003*	0.007	0.978	0.007	0.007	*1.2*	1.4	58.7	1.4	1.4	*1*	1	13	1	1	64.0287
14	*0.003*	0.007	0.007	0.978	0.007	*1.2*	1.4	1.4	58.7	1.4	*1*	1	1	13	1	64.0287
15	*0.003*	0.007	0.007	0.007	0.978	*1.2*	1.4	1.4	1.4	58.7	*1*	1	1	1	13	64.0287
16	*0.014*	0.043	0.314	0.314	0.314	*2.0*	3.5	19.6	19.6	19.6	3	4	11	11	11	64.1379
17	*0.014*	0.314	0.043	0.314	0.314	*2.0*	19.6	3.5	19.6	19.6	3	11	4	11	11	64.1379
18	*0.014*	0.314	0.314	0.043	0.314	*2.0*	19.6	19.6	3.5	19.6	3	11	11	4	11	64.1379
19	*0.014*	0.314	0.314	0.314	0.043	*2.0*	19.6	19.6	19.6	3.5	3	11	11	11	4	64.1379
20	*0.009*	0.473	0.473	0.022	0.022	*1.6*	28.9	28.9	2.3	2.3	2	12	12	3	3	64.0855
21	*0.009*	0.473	0.022	0.473	0.022	*1.6*	28.9	2.3	28.9	2.3	2	12	3	12	3	64.0855
22	*0.009*	0.473	0.022	0.022	0.473	*1.6*	28.9	2.3	2.3	28.9	2	12	3	3	12	64.0855
23	*0.009*	0.022	0.473	0.473	0.022	*1.6*	2.3	28.9	28.9	2.3	2	3	12	12	3	64.0855
24	*0.009*	0.022	0.473	0.022	0.473	*1.6*	2.3	28.9	2.3	28.9	2	3	12	3	12	64.0855
25	*0.009*	0.022	0.022	0.473	0.473	*1.6*	2.3	2.3	28.9	28.9	2	3	3	12	12	64.0855

Source: See text. Values for agent 1 are italicized to indicate that this agent's endowment is different from that of the other agents.

Table 2.3 Equilibrium prices, incomes, and utility rankings, 5 × 5 economy, asymmetrical endowments and unequal substitution elasticities

Equilibrium	Prices					Incomes					Utility rankings (5 = best)					GDP
	p_1	p_2	p_3	p_4	p_5	y_1	y_2	y_3	y_4	y_5	u_1	u_2	u_3	u_4	u_5	
1	0.692	0.275	0.016	0.012	0.006	48.7	17.2	1.9	1.7	1.4	4	4	3	5	5	71.0
2	0.962	0.020	0.009	0.006	0.003	67.4	2.2	1.5	1.4	1.2	5	1	2	3	4	73.7
3	0.005	0.985	0.005	0.003	0.002	1.3	59.1	1.3	1.2	1.1	1	5	1	1	1	64.07
4	0.007	0.021	0.963	0.005	0.003	1.5	2.3	57.8	1.3	1.2	2	2	5	2	2	64.10
5	0.011	0.159	0.816	0.009	0.004	1.8	10.4	49.1	1.6	1.3	3	3	4	4	3	64.2
.	.	.	.	.	.	.	.	.	.	.	.	.	.	.	.	.
.	.	.	.	.	.	.	.	.	.	.	.	.	.	.	.	.
.	.	.	.	.	.	.	.	.	.	.	.	.	.	.	.	.
?																

Source: See text.

associated with equilibrium 2 is highest, yet a majority (in fact, all but agent 1) would prefer equilibrium 1 over equilibrium 2. As in the previous cases, some of these equilibria are characterized by a highly unequal distribution of income; this results when the price of one particular good is considerably higher than the price of the other goods. (If an individual is endowed with a disproportionately large share of a high-priced good, that individual's income will be disproportionately high in the equilibrium.) The model differs from the other two in that an agent's personal income does not give an unambiguous indication of the individual's utility ranking of an equilibrium. Agent 5's income is higher in equilibrium 5 than in equilibrium 2, yet this agent's utility is greater in equilibrium 2.

Different multiple equilibria can arise under other variations in specification of the model. As the parameters and endowments vary, there can be different numbers of equilibria.[33] For some ranges of the parameters the equilibrium is unique, although a full search of the parameter space to find the region having a unique equilibrium is impractical even in the case of these very simple 5×5 models. The dimensionality of the problem is just too large. With n commodities, each agent would require specification of $n - 1$ utility weights a_{ij} (each agent's utility function can be rescaled to make $a_{i1} = 1$), one substitutability parameter b_i, and n endowments ω_{ij}, or $2n$ parameters in all. Thus, with n agents and n goods the full parameter space would have $n \times (2n) = 2n^2$ parameters. Defining the commodity units by setting the ω_{ij} for one agent all equal to one would reduce the number of parameters to $2n^2 - n$. If the substitution parameters differed across goods for each consumer (so that there were n of the b's for each consumer), the total number of parameters would be $3n^2 - 2n$. Thus, the number of parameters increases quadratically with the size of the economy. The ranges of these parameters are bounded from only one side – the a_{ij}'s can be presumed to be positive, the b_i's less than one, and the ω_{ij}'s greater than zero – so there is no way to search the parameter space over a finite grid. And this is for only a simple exchange economy; the number of parameters increases even more if production were to be introduced.

2.3.2 The unknown dynamics of the general equilibrium system

The equilibrium prices and incomes shown in Tables 2.1–2.3 were calculated using numerical routines for solving a simultaneous system of nonlinear equations.[34] This does not mean that a real *economy* would be able to find or would converge to any of these equilibria; that would depend on the *actual dynamics* of the economic system. Economists have shied away from incorporating dynamic adjustment mechanisms into their integrated assessment models for good reasons.[35] If transactions are allowed to take place "out of equilibrium," the model would have to deal with the possibility of shortages and surpluses – and the plans of consumers and producers would not be fulfilled. Nonfulfillment of plans means that the maximization that presumably defines the behavior of the producers and consumers is incomplete. This would destroy any favorable welfare interpretation of the prices and transacted quantities observed in the markets. But the theoretical difficulties run deeper; transactions "out of equilibrium" would also affect the wealth of the agents, and would therefore change the conditions for their maximization subsequently in unforeseeable ways (Fisher 1983, 1989).[36] On the other hand, if out of equilibrium transactions are to be avoided, there must be some mechanism for arriving at the equilibrium prices prior to any transactions taking place. Obviously, no such economy-wide mechanism exists.

This lack of a well-defined equilibrium-finding mechanism has been an embarrassment since the earliest days of mathematical economics and general equilibrium theory. The most familiar way out is the tâtonnement ("groping") process of the type originally hypothesized by Walras. In the tâtonnement, it is imagined that prices adjust according to excess demands – if there is positive excess demand for a commodity, its price increases, while if excess demand is negative, the price decreases.[37] In elementary treatments, something like the adjustment process situation depicted in Figure 2.4 is presumed to be taking place.

Figure 2.4 represents a particular market, say for automobiles. The demand and supply curves are shown as depending on the price of automobiles (although note that in the background, the level of income of consumers is an important determinant of demand). If for any particular price of automobiles P_0, the amount that auto manufacturers would like to sell (given by the supply curve) is greater than the amount consumers want to buy (indicated by the demand curve), there will be excess supply. This could manifest itself as a buildup of inventories, or a slowdown in the rate of sales from what is "normal." Faced with these conditions, suppliers would cut prices with the aim of working down the inventories and stimulating demand. Similarly, if P_0' were "too low" (i.e., below equilibrium) consumers would want to buy more cars than producers were willing to sell, and the situation would be one of

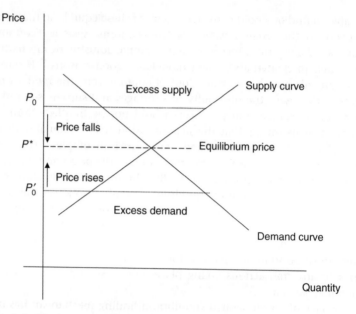

Figure 2.4 Moving to equilibrium

excess demand. Inventories would be falling, consumers would face a shortage of new cars, and suppliers would adjust prices upward. In either case, the price adjustment process would settle down only when the price reached the level P^*, the equilibrium price where supply and demand are equal (or where excess demand is zero). This, in short, is the version of the tâtonnement process found in elementary economics textbooks. It underlies popular notions of "the laws of supply and demand" that presumably regulate the market system.

This story is an oversimplification that leaves out essential elements, however. In a general equilibrium system that represents the activity of the entire economy, markets are not isolated from each other. They are interconnected both through the influence that the price of one good has on the demand for another (for example, the price of gasoline influences the demand for automobiles), and because prices affect the income and wealth of the consumers (recall that the prices of other goods and consumers' incomes were in the background of Figure 2.4, i.e., they were not treated explicitly). These interdependencies are represented by the fact that the excess demand function of each commodity (commodity j for example) depends on *all* the prices (as well as the endowments and other parameters), not just the price of commodity j. Therefore, the appropriate mathematical representation of the tâtonnement is something like the equations (2.6), a set of time differential equations for each price p_j where the excess demands f^j are those of equation (2.5) and the k_j are "speed of adjustment" parameters.

$$\frac{dp_j}{dt} = k_j f^j(p_1, \ldots, p_n; \forall a_{ij}, \omega_{ij}, b_i), \quad j = 1, 2, \ldots, n \tag{2.6}$$

The tâtonnement process is *less general* than the numerical methods used by Mathematica (and other software) to solve systems of equations, because in the tâtonnement, the change in a commodity's price depends only on its own excess demand function. Economists have not been successful in telling stories of how market prices might adjust more generally.[38]

From the equations in (2.6), it can be seen that the features of the tâtonnement equations depend entirely on the properties of the excess demand equations (the f^j). But because the excess demand equations can have virtually any shape, the tâtonnement trajectories can aim in any direction. Hence there is no guarantee that the tâtonnement will converge to any particular equilibrium (Mas-Colell et al. 1995, pp. 621–3).

In the models of Tables 2.1–2.3, the tâtonnement process is not capable of finding all the equilibria, no matter what its starting point (the initial values of p_1, p_2, p_3, p_4, and p_5) might be. For example, in the symmetrical model of Table 2.1 the only equilibria that can be found via tâtonnement are 6–10. An equilibrium can be said to be "stable" if it will be reached by the tâtonnement process. It is known that an equilibrium is stable if the goods are "gross substitutes" at the point of equilibrium. The property of gross substitutability is that "if one commodity price goes up while all other prices remain unchanged, there will be an increase in [market] excess demand for every commodity whose price has remained constant" (Arrow et al. 1959, p. 86). This condition can be checked by examining the matrix of partial derivatives of the excess demand functions with respect to the individual prices, evaluated at the equilibrium. Computation of the matrix at equilibria 1–5 and 11–31 of Table 2.1 shows that none of them is stable. Interestingly, in the asymmetrical model of Table 2.3, the three equilibria with highly unequal income distributions (2, 3, and 4) are stable under the tâtonnement process, while equilibria 1 and 5 are not. In these cases, the stable equilibria are those which are best for a single agent, and worst for everyone else.

Knowledge of the dynamics of the economic system is, nevertheless, an essential component of understanding how the system would react to any kind of shock, including a policy change. It was proved by Debreu (1970) that economies of the type being considered here are "regular."[39] What happens if a "regular" economy is perturbed from one of its finite number of equilibria? The dynamic path followed by the economy will determine whether it returns to that equilibrium or to another one. Some kinds of dynamics might not enable the economy to reach equilibria that might be, on social or political grounds, desirable. The *actual* outcome of any "comparative statics" exercise involving a policy change or parametric shock will also depend on the *real dynamics of the economy*. Given the possibility of multiple equilibria, an economic model, to be useful for policy analysis purposes, must include a specification of economic dynamics.

2.3.3 Can current models reveal multiple equlibria and unstable dynamics?

In the empirical implementation of general equilibrium modeling, there are several ways to arrive at the representation of consumers and their preferences. The first is to specify market demand functions and estimate them from aggregate data, relying on the interaction of these demand functions and production-based supply functions (perhaps also estimated from aggregate data) to determine prices and outputs in the general equilibrium solution of the model. Often, these market demand functions are presumed to represent the demand of a "representative consumer" or consumers who exhibit the same characteristics as real individuals. A second approach is to start with a specification of the consumers' utility functions, and to estimate a "demand system" based on price and expenditure data. The market demand equations would then be built up by summing the individuals' demand functions. A third approach is to specify demand functions a priori, sometimes relying on studies in the literature to provide values of the relevant parameters. The functions may be calibrated to aggregate data, or some combination of calibration and econometric estimation may be used. In practice, each of these methods has drawbacks. None offers a fully satisfactory answer to the question of whether or not multiple equilibria are possible.

Consider first the specification of market demand functions based on the "representative agent" approach. It has been observed by no less an authority than James Tobin that an economics paper that does not begin with rational maximization by representative agents (i.e., that is not based on "microfoundations") would be unlikely to see the light of day:

> [The microfoundations] counter-revolution has swept the profession until now it is scarcely an exaggeration to say that no paper that does not employ the "microfoundations" methodology can get published in a major professional journal, that no research proposal that is suspect of violating its precept can survive peer review, that no newly minted Ph.D. who can't show that his hypothesized behavioral relations are properly derived can get a good academic job.
>
> (Tobin 1986, quoted in Rizvi 1994, p. 350)

It is hard to understand why this self-imposed methodological constraint is so uniformly upheld. It has been known for some time that the representative agent approach is fundamentally inconsistent. The reason is that if a representative agent is constructed on the basis that this agent reproduces market outcomes, the implied preferences of the representative agent can be in conflict with the actual preferences of the agents making up the economy (Jerison 1984, cited in Kirman 1992). The representative agent may prefer commodity bundle A to bundle B at some set of prices, even though the individual agents in the economy all prefer B to A. Once again, individual demands do not aggregate to a market demand of the same form. Kirman's appraisal is telling:

> A tentative conclusion, at this point, would be that the representative agent approach is fatally flawed because it attempts to impose order on the economy through the concept of an omniscient individual. In reality, individuals operate in very small subsets of the economy and interact with those with whom they have dealings. . . .
>
> (1992, p. 132)

Kirman goes on to describe some promising attempts that have been made to show how *systematic* outcomes could result from such localized interactions, but he concludes that

> [t]he equilibria of the worlds described by any of these approaches may be conceptually very different from those implied by the

artifact of the representative individual. . . . In a world with many interacting heterogeneous agents, a natural idea of an equilibrium would be not a particular state, but rather a distribution over states, reflecting the proportion of time the economy spends in each of the states. This distribution could exhibit very regular characteristics, while remaining far from reflecting the behavior of a single maximizing agent.

(1992, p. 133, footnote and references omitted)

Kirman's judgment at the beginning of his review is a fair summary: "[T]he 'representative' agent deserves a decent burial, as an approach to economic analysis that is not only primitive, but fundamentally erroneous" (1992, p. 119).

Building up aggregate demand functions out of demand systems estimated from individual expenditure data also faces difficulties. If important elements of the utility function, such as the environment, are not priced, then in practice no data on consumer expenditures can reveal the parameters of the utility function that are associated with that "commodity." This is precisely the situation with regard to climate stability and other desirable attributes of the global environment.[40] Even for the commodities that are traded in markets, severe dimensionality problems prevent the recovery of all of the parameters of the consumers' utility functions. Flexible functional forms derived as Taylor series approximations of general utility functions have a number of parameters that increases quadratically with the number of goods. The data requirements for estimating such a demand system are formidable, and as a result such systems are typically estimated only after a number of restrictive assumptions are made that drastically reduce the number of free parameters. Examples of such restrictions include specifying that some of the parameters are the same across all consumers, or that certain elasticities are known a priori. This brings us back to the problems associated with positing a "representative consumer" and with limiting the potential complementarity of goods.

The third alternative is simply to rely on aggregate statistical information to estimate market demand and supply curves for the various commodities traded in the economy. Such a rough-and-ready approach has some appeal, except that the results of the estimation procedure cannot then be used to make welfare arguments. If the empirical demand and supply curves are robust under alternative policy scenarios (this is itself a questionable assumption), then the empirically based estimates might have some predictive value. However, hypothetical changes in market aggregates calculated on the basis of such a model (such as changes in GDP or prices) would have no normative implications, because they could not be mapped to changes in the circumstances or preferences of individuals.

Another very serious problem with atheoretical estimates has to do with specification of their functional forms for estimation. We know from the Sonnenschein–Mantel–Debreu theorem that the market demand functions can have almost any shape. If so, what functional forms would be appropriate for estimation of those functions based on market data? It would not be possible to appeal to the theory of the individual consumer for restrictions; the point of the theoretical results is that the aggregate demand functions do *not* necessarily have the same properties as individual demand functions. Yet some restrictions are necessary both to identify the functions (Fisher 1966) and to produce sufficient degrees of freedom for statistical estimation. It is difficult to imagine a credible procedure for atheoretical estimation in the face of fundamental indeterminacy of the underlying functional forms.

Most common in the practice of integrated assessment is to assume that the utility functions of the consumers take on some highly simplified form. Usually representative consumers are specified; the parameters are selected by the analyst (with perhaps some reference to empirical work). Collapsing preferences into this kind of representation rules out by assumption the problems of multiple equilibria and instability if the posited utility functions impose a high degree of substitutability on the different goods (or if the differences in goods are simply assumed away and utility is specified to depend on aggregate GDP). For example, the Cobb–Douglas utility function implies that the elasticity of substitution between all goods is unity. It amounts to imposing gross substitutability (which, in exchange models, implies uniqueness of equilibrium and tâtonnement stability) by *assumption*.

To provide a flavor of the current "state of the art" in integrated assessment modeling, consider the sample of models that participated in the recent Energy Modeling Forum evaluation of the "costs of Kyoto." A special issue of the *Energy Journal* compiled the results of the runs of 13 models, along with descriptions of the scenarios that were simulated. The modelers gave summary descriptions of their models and how they implemented the scenarios. How are consumers' preferences represented in these models?

Table 2.4 shows how each model does it. Recall that logarithmic utility functions are the same as Cobb–Douglas, because any monotonic

Table 2.4   Representation of preferences in integrated assessment models

Model	Representative consumers (for regions)	Intraperiod functional form of utility function	Interperiod functional form of utility function
MERGE[a]	yes	logarithmic	discounted sum
SGM[b]	yes	demand[b]	demand[b]
EPPA[c]	yes	Cobb–Douglas	single–period
RICE[d]	yes	logarithmic	discounted sum
FUND[e]	yes	adjusted log	discounted sum
GRAPE[f]	yes	logarithmic[f]	discounted sum
WorldScan[g]	yes	Cobb–Douglas[g]	discounted sum[g]
AIM[h]	yes[h]	_[h]	_[h]
MS-MRT[i]	yes	Cobb–Douglas[i]	discounted sum CES[i]
GTEM[j]	yes	CDE[j]	single-period[j]
G-Cubed[k]	yes	Cobb–Douglas[k]	discounted sum log
Oxford[l]	–	–	–
CETA-M[m]	yes	logarithmic	discounted sum

Notes: The papers in the special issue of *The Energy Journal* are Manne and Richels (1999), MacCracken et al. (1999), Jacoby and Wing (1999), Nordhaus and Boyer (1999), Tol (1999), Kurosawa et al. (1999), Bollen et al. (1999), Kainuma et al. (1999), Bernstein et al. (1999b), Tulpulé et al. (1999), McKibbin et al. (1999), Cooper et al. (1999), and Peck and Teisberg (1999).

[a] MERGE: "Global Negishi welfare is a weighted sum of the discounted logarithm of macroeconomic consumption, adjusted for the economic losses generated by non-market damages. . . . In the Pareto-optimal case, nonmarket damages enter the Negishi welfare definition, and market damages represent one of the competing claims on the allocation of total production resources. In all other cases, nonmarket and market damages are excluded from the determination of the equilibrium, but are determined in a post-optimization step. In this way, we allow for the possibility of non-cooperation or of non-optimal international control agreements." (For further discussion, see the website http://www.stanford.edu/group/MERGE/.)

[b] SGM: There are four final demand sectors (personal consumption, government consumption, investment, and net exports). Household utility functions are not defined; instead, household demand for each final product is specified as a normalized demand function with a price and income elasticity. The household supplies of labor and savings are also modeled as supply functions and are not derived from underlying utility functions. Government utility is CES in general government services, national defense, and education (Sands et al. 1999).

[c] EPPA: The representative agent in each region for each period maximizes $U = \Pi_c (Y_c - \theta_c)^{\mu c}$ for the four consumption goods indexed by c, with Y_c equal to consumption and θ_c a minimum consumption level (Yang et al. 1996). This modified Cobb–Douglas utility function is also referred to as a "linear expenditure system." Utility is maximized in each period, and "[f]actor endowments are updated at each step, according to assumed exogenous trends in rates of population growth, increases in labor productivity, autonomous energy efficiency improvement (AEEI), and availability of natural resources" (Jacoby and Wing 1999, p. 77, citing Yang et al. 1996).

Notes to Table 2.4 continued

d RICE: The RICE model notionally specifies an additively separable (for each time period) CES utility function with consumption per capita in each period as the arguments, but when it comes to carrying out the actual runs of the model, the substitution parameter is chosen so as to make the utility function logarithmic (Nordhaus and Boyer 2000).

e FUND: Utility is the logarithm of per capita consumption minus investment minus market damages, plus a second-order Taylor approximation for monetized nonmarket damages. The Taylor approximation is used because "in some uncertainty analyses so many people died that the non-market losses exceed income" (Tol, personal communication, 2001).

f GRAPE: A single utility function consisting of a weighted (with Negishi weights) sum of the individual regions' utilities functions is maximized.

g WorldScan: Intertemporal optimization by the representative agents maximizes the discounted sum of the logarithm of annual consumption, where the discount rate includes the probability that the consumer will be alive in the next period. The instantaneous utility functions are Cobb–Douglas in the seven categories of goods. Demand for a particular category within a region is a CES composite of all varieties produced in the different regions (CPB Netherlands Bureau of Economic Policy Analysis 1999).

h AIM: Model structure not accessible.

i MS-MRT: The nonenergy goods enter the intraperiod utility function in Cobb–Douglas form, but end-use energy consumption enters the utility function directly in CES form with the aggregate of nonenergy goods. The intertemporal sum that is maximized is a separable utility function with a constant elasticity of intertemporal substitution. Imports from different regions and domestically produced goods are distinguished (Bernstein et al. 1999b).

j GTEM: The commodities include coal, oil, and natural gas. Intraperiod utility functions are of the "constant difference elasticity" (CDE) type (Hanoch 1975). Income in a given period is divided between savings and consumption in fixes proportions according to a life-cycle model and demographic representation of the composition of the population. Additional dynamic equations represent adjustment of stock variables (ABARE 2002).

k G-Cubed: The intertemporal utility function that is maximized is of the form

$$U_t = \int_t^{\infty} [\ln C(s) + \ln G(s)] e^{-\theta(s-t)} ds$$

where $C(s)$ is the household's aggregate consumption of goods at time s, $G(s)$ is government consumption, and θ is the rate of time preference (set equal to 2.5%). The final consumption function is modified to account for deviations from the permanent income hypothesis. Within each period, $C(s)$ is treated as a nested CES function; the top tier consists of inputs of capital services, labor, energy, and materials; energy and materials are CES aggregates of inputs of individual goods. "The elasticities of substitution at the energy and materials tiers were estimated to be 0.8 and 1.0, respectively. In this version of the model the top tier elasticity has been imposed to be unity" (McKibbin et al. 1999, pp. 295–6). Thus, at all levels of the utility function except the energy aggregate, the specification is Cobb–Douglas (elasticity of substitution of unity).

l The Oxford model is not a general equilibrium model and does not contain an explicit representation of preferences.

m CETA-M: Market and nonmarket damages monetized. See Peck and Teisberg (1992) for specification of the utility function.

transformation of a utility function has the same observable implications as the original utility function. Thus, we can see that in all but
one of the models participating in EMF-16, either the intraperiod utility
functions were Cobb–Douglas and did *not* include environmental goods,
or the value of environmental goods was simply monetized. In the latter
case, of course, the implicit elasticity of substitution between the environmental goods and other goods is infinite – they are perfectly substitutable at the prices chosen for the monetization. For Cobb–Douglas
utility functions defined over ordinary goods, the market demand
functions are *linear* in normalized (or relative) prices, and hence the
equilibrium necessarily is unique. Specification of Cobb–Douglas (or
logarithmic) utility rules out by assumption the possibility of multiple
equilibria originating on the demand side. In all but one of the models,
the intertemporal substitutability of the goods was also high by definition. The implications of that assumption will be treated in the following chapter.

2.4 Conclusions and implications for policy

What can we conclude from this review of the treatment of preferences
in integrated assessment models? It is clear that the existing models are
set up in such a way as to preclude any of the problems with multiplicity of equilibria or instability of equilibria arising from the demand
side. This is not to say that the real world does in fact exhibit multiplicity of equilibria and instability; that is a question that remains to be
determined. A great deal of empirical work would be required, especially
because the representation of nonmarketed environmental goods in
individual preferences is a subject that is largely unexplored.

The inadequacies of the representation of consumers in standard
economic climate policy models can therefore be summarized as
follows:

- The "outside" critique of preference representation is entirely
  ignored;
- For intraperiod utility, the diversity of individual consumers within
  each region of analysis (typically a large country such as the US or
  China, or a multicountry region such as the EU) is replaced by a
  representative agent, even though it is known that the representative agent assumption is not adequate to capture the actual shape of
  market demand functions;
- Either environmental goods are left out of the utility function, or

their contribution to well-being is simply monetized, thereby precluding investigation of any questions having to do with the potential complementarity of environmental and nonenvironmental goods. If environmental goods are monetized (as in conventional CBA), the implicit assumption is that the elasticity of substitution for nonmarginal changes in the composition of environmental and nonenvironmental goods in utility is infinite;

- The preferences of the representative agents are typically described by logarithmic or Cobb–Douglas utility functions, thereby imposing gross substitutability. Multiple or unstable equilibria arising out of the properties of consumers' demand functions are thereby ruled out.

Current specimens of climate/economy applied general equilibrium models thus are *incapable of revealing multiple equilibria, even if the real economy has them.* The project of establishing such models on "microfoundations" is not a movement in the direction of greater rigor; instead, it creates the erroneous impression that modeling conclusions have been derived from unassailable scientific procedures, when in fact the results are no more than the consequences of the assumptions built into the models. Very strong (and entirely untested) assumptions are required to rule out the possibilities of multiple equilibria and dynamic instability. In reality, the true number of equilibria in the real economy is not known. Nor is very much known about the actual process of dynamic adjustment within the economy. What is known is that even elementary general equilibrium systems can exhibit diverse and problematic phenomena. The multiple equilibria of the simple exchange models discussed here should give pause to those who would unquestioningly accept the results of the current generation of more complicated integrated assessment models.

3
The Treatment of Time

3.1 The problem

Climate change takes place over decades, centuries, and millennia. The consequences of actions taken now also work themselves out on time scales covering multiple generations. The persistence of CO_2 in the atmosphere depends on the long-term response of the biosphere to increased temperatures and concentrations, and the atmospheric lifetimes of many of the non-CO_2 greenhouse gases are over 100 years. The achievement of thermal equilibrium between the surface layers of the ocean and atmosphere takes decades, and achievement of equilibrium between the different layers of the ocean requires even more time (IPCC 2001a).[1] These physical facts mean that the treatment of events and policies that unfold over *very long stretches of time* must be a central feature of the economic analysis of climate change. Not only do policies have long-lasting consequences, but the basic conceptualization of the economic system itself has to reflect the lengths of time involved.

To do this appropriately is no trivial problem. Standard introductory textbook treatments of time are inadequate. Nothing illustrates this better than the "paradox of discounting." If the future benefits and costs of a climate management policy are discounted at normal market rates, say 7 percent per annum, then nothing that happens in the distant future can matter for the decisions taken today. For example, suppose that the discount rate is 7 percent and world GDP grows at 2 percent per annum. Then the present value of the entire world GDP starting at a point 300 years in the future *and going on forever* is only about $374 million, or roughly 6¢ per capita.[2] Even if only half the people now alive cared about the fate of civilization 300 years from now, the "cost" of such an economic apocalypse (assuming the validity of the present

58

value calculation) would be only 12¢ per capita to those who care. The conventional interpretation of this calculation is that it would be worth very little to people today to avert the loss of all economic output from 300 years in the future until the end of time. Yet such a loss would surely mean the permanent end of civilization as we know it (albeit 300 years in the future), and subjectively at least, we *know* that people today would be willing to invest substantial resources today to avert such an outcome.

This is only part of the conundrum, however. From the conventional point of view, it is equally unappealing to assert that the discount rate "should" be zero for purposes of assessing the value of costs and benefits accruing far in the future. If the discount rate were zero, then even the tiniest permanent increase in future consumption would have an indefinitely large – actually infinite – present value. Thus, any sacrifice of consumption in the present, no matter how large, would be justified if it could produce a small permanent increase in future consumption. As long as investment has a positive payoff in productivity, such a trade-off (of present consumption for future consumption) is possible. Subjectively, however, the present generation would no more be willing to reduce its standard of living to subsistence (or below) in order to increase future generations' consumption by a trivial amount than it would be willing to condemn the future generations to extinction in return for minor present benefits. Thus, discounting by any positive factor would seem to lead the present generation to disregard the fate of the future generations, while zero discounting implies that the interests of future generations completely outweigh that of the present. Hence the paradox.[3]

The problem has to do with how intertemporal comparisons of costs and benefits are made. While it is easy to construct justifications for using a market-based discount rate to compare transfers of consumption between different points in time for an individual, the balance of empirical research suggests that even for individuals the simple "discounted utility" model is not consistent with the evidence.[4] In the conclusion of their review article on the subject, Frederick et al. state that

> The DU [discounted utility] model, which continues to be widely used by economists, has little empirical support. Even its developers – Samuelson, who originally proposed the model, and Koopmans, who provided the first axiomatic derivation – had concerns about its descriptive realism, and it was never empirically validated as the

appropriate model for intertemporal choice. Indeed, virtually every core and ancillary assumption of the DU model has been called into question by empirical evidence collected in the past two decades. . . .

While the DU model assumes that intertemporal preferences can be characterized by a single discount rate, the large empirical literature devoted to measuring discount rates has failed to establish any stable estimate. There is extraordinary variation across studies, and sometimes even within studies. . . . [T]he spectacular cross-study differences in discount rates also reflect the diversity of considerations that are relevant in intertemporal choices and that legitimately affect different types of intertemporal choices differently. Thus, there is no reason to expect that discount rates *should* be consistent across different choices.[5]

(2002, p. 393, emphasis in the original)

Aside from problems with the discounting model as applied to *individuals'* intertemporal choices, the time scales of the climate problem add additional complications. Climate issues span multiple generations, not just the lifetimes of people who are alive today, so proper modeling requires a suitable treatment of the fact that the economy lasts longer than the lifetimes of the individuals who are alive at any particular moment. In formulating climate policies, the interests and well-being of people who are not yet alive are at stake. These future individuals do not participate in present-day decision-making, whether political or economic. These facts underlie why the mechanical application of the discount formula fails as a climate modeling strategy, even if the DU model could validly be applied to individuals.[6] This chapter will discuss some of the methods economists use to model activity stretching out over very long time periods. As we shall see, the same kinds of problems with multiple equilibria, indeterminacy, and instability that plague static general equilibrium models are reproduced, with even greater severity, in long-period models.

The previous chapter dealt with static general equilibrium models where consumers are diverse and goods are complementary. Under these circumstances, wealth effects stemming from different endowments can be important, leading to the possibility of multiple equilibria and unstable dynamics. The unsolved problem of "equilibrium choice" when there is more than one Pareto-optimal competitive outcome looms large. We also saw that the specification of preferences in most of the leading integrated assessment models rules out any of these complications by assumption. The result of such a restriction on the way preferences are represented means that the conclusions of the analysis are dependent in a crucial but nontransparent way on the premises built into the models.

The type of model required to address global climate change cannot be a static one, however. Because of the long time periods involved, economic analyses of climate policy must incorporate the time dimension in an *essential* way. We will see that this requirement exacerbates the kinds of unsolved problems that plague static models, no matter what approach to the treatment of time is chosen by the modeler. The issues of multiple equilibria and dubious dynamics are endemic to these models. While in static models, there are generally a finite number of equilibria, in some dynamic models there are an infinite number of possible equilibrium paths. Just as in the static models, the properties of equilibria depend on the endowments of the various forms of wealth possessed by the agents, but in intertemporal models the pattern of those allocations across individuals is even more starkly dependent on policy choices than in the static case. Finally, in some intertemporal models, the equilibria depend crucially on *expectations about the future*, and these expectations cannot in any reasonable way be determined by the economic "fundamentals." The expectations-formation mechanism therefore provides yet another source of indeterminacy in climate policy models.

The chapter will proceed as follows. First will come a discussion of the three main ways intertemporal models can be set up:

1. The "useful fiction" that all transactions occur at the beginning of time;
2. Models in which markets clear in successive time periods, based on historical experience and expectations about the future;
3. "Social planner" models in which an imaginary collective decision-maker maximizes the utility of society's members over an infinite horizon.

The plan will be to introduce the sparsest models first, adding features that allow additional phenomena to emerge. As throughout this book, the modeling approach will emphasize simplicity and transparency. The models introduced will show some of the things that can happen in intertemporal models, and why climate/economy models would have to address a much broader range of issues than they now do if they were to provide useful guidance for policy.

3.2 Three ways to represent time

There are basically three ways in which intertemporal general equilibrium models can be specified. The first is grounded in the Arrow–Debreu tradition, and imagines that all transactions take place at a single (initial) moment in time, for goods that "occur" at various (present and future) points in time. The intertemporal nature of the model is achieved by dating the goods, so that the same physical commodity transacted today and tomorrow becomes two different commodities distinguished by their dates.[7] Conceptually, this "useful fiction" (of all transactions being settled once and for all at the outset) collapses the model to the same mathematical structure as the static models examined in the previous chapter, with the possible exception that there will be an infinite number of goods if the economy is assumed to extend indefinitely into the future. Such models have been developed, and they exhibit the same features as finite-dimensional models (see section 3.5). In fact, the essential characteristics for our purposes can be described by a two-period model. The reason is that the agents in the two-period model can be interpreted as corresponding to people who would be alive at different times, so that their preferences reflect their actual circumstances. We will see that any intertemporal model that takes the "useful fiction" of one-time transacting seriously has to specify agents who are diverse in precisely the ways that lead to multiple equilibria and unstable dynamics.

The second way time can be represented in integrated assessment models portrays a sequence of market-clearing equilibria occurring in successive time periods. This class of models is best represented by the "overlapping generations" (OLG) framework, in which people who are born at different times transact with each other because their lifetimes overlap. Each period's equilibrium is determined by the "fundamentals" (the preferences and endowments of the individuals alive in that period), but depends as well on past history (because past equilibrium prices can affect the wealth of people alive today who also transacted in the previous period) and on expectations about the future (because the allocations that satisfy individuals' maximization conditions will reflect expected values of future prices). These OLG models will result in equilibria that depend on the endowments possessed by individuals, but will have to include as well some kind of mechanism for *expectations formation.* We will see that this feature gives rise to additional sources of multiplicity.

The third approach involves an imaginary "social planner" maximizing the utilities of society's members over an infinite horizon. Alternatively, the different agents can be thought of as "dynasties" that develop rules for consumption based on maximization of a utility function that spans infinite time. Such models are much less satisfactory than either the Arrow–Debreu simultaneous transactions models or the OLG models, because in addition to all the other abstractions inherent in the modeling exercise, social planner models are based on the action of a fictitious benevolent despot who does not and cannot exist. Even if some rule of behavior could be derived that would solve the maximization problem, it might be in the interest of subsequent members of society to change it, so there is no way to guarantee that it would be adhered to over time. In addition, by vesting all power in the hands of a social planning agency, this type of modeling fosters the political attitudes that lead to a "dictatorship of the present."[8] We shall see that this point of view is at the heart of the biases inherent in current integrated assessment modeling practice. In addition, there are technical economic and mathematical reasons why this kind of treatment of time is less satisfactory than either of the other two approaches.

3.3 The Arrow–Debreu approach: all transactions occur at the "beginning"

3.3.1 One agent, one good, two time periods: the simplest case

First consider the most basic kind of "one-time" transactions intertemporal model, with a single agent, one good, and two time periods. The preferences of this agent are represented by a stripped-down CES utility function,

$$u = a_1 \frac{x_t^b}{b} + a_2 \frac{x_{t+1}^b}{b} \tag{3.1}$$

where x_t and x_{t+1} are the amounts of the good consumed in the two periods, b is the substitutability parameter, and a_1 and a_2 are utility weights. The agent's endowments of the good in the two periods are ω_t and ω_{t+1}. The agent maximizes utility subject to the full-duration budget constraint

$$W = p_t\omega_t + p_{t+1}\omega_{t+1} - p_t x_t - p_{t+1}x_{t+1} \qquad (3.2)$$

where p_t and p_{t+1} are the prices of the good in each of the two periods, respectively. This maximization gives rise to a demand function in each period, namely

$$x_t^* = \frac{\left(\dfrac{a_2 p_t}{a_1 p_{t+1}}\right)^{\frac{1}{b-1}}(p_t\omega_t + p_{t+1}\omega_{t+1})}{p_t\left(\dfrac{a_2 p_t}{a_1 p_{t+1}}\right)^{\frac{1}{b-1}} + p_{t+1}} \qquad (3.3)$$

and

$$x_{t+1}^* = \frac{(p_t\omega_t + p_{t+1}\omega_{t+1})}{p_t\left(\dfrac{a_2 p_t}{a_1 p_{t+1}}\right)^{\frac{1}{b-1}} + p_{t+1}} \qquad (3.4)$$

As in all such general equilibrium models, a price normalization is required. For reasons that will become apparent momentarily, the normalization $\beta = p_{t+1}/p_t$ will be used. The ratio β will be referred to as the "inflation ratio." (The ordinary "rate of inflation" is $1 - \beta$.) If the good cannot be stored so that excess demand (defined as the difference between the utility-maximizing quantity demanded and the endowment) must be zero in each period, the equilibrium solution is

$$\beta^* = \left(\frac{a_2}{a_1}\right)\left(\frac{\omega_t}{\omega_{t+1}}\right)^{1-b} \qquad (3.5)$$

Because the commodity cannot be stored, consumption in each period is just equal to the endowment and in some sense this "equilibrium" is trivial – no alternative allocations of consumption across periods are possible. Nevertheless, the equilibrium concept is useful because it shows what prices (or price ratio β^*) is consistent with maximization of the agent's preferences given his utility function and the values of the endowments. As is apparent from equation (3.5), the equilibrium β^* is a function of the ratio of the endowments (ω_t/ω_{t+1}) and the utility weights (a_2/a_1), and also depends on the substitutability parameter b.

As simple and seemingly trivial as it is, this model embodies the essence of the treatment of time and preferences common to the current generation of integrated assessment models! To see why this is true, we need to translate the notation of equation (3.5) into a more familiar form. The endowment ratio (ω_t/ω_{t+1}) corresponds to the growth of the economy, the utility weight ratio embodies the subjective rate of time preference, and the equilibrium intertemporal price ratio can be straightforwardly transformed into the (endogenous) market rate of interest. The substitutability parameter b determines both the intertemporal elasticity of substitution and the "elasticity of the marginal utility of consumption." Once these translations are made, it will be seen that equation (3.5) is nothing but the formula for the "descriptive" value of the discount rate that is typically used in standard integrated assessment models (Cline 1992, Nordhaus 1994, IPCC 1996).

First, in what sense is β just another way of expressing the market rate of interest? Let r be the interest rate as conventionally defined. Then one unit of the good in the future period is worth $1/(1 + r)$ units of the good today. But a unit of the good in the future period is worth p_{t+1}, and a unit of the good today is worth p_t. Hence $p_{t+1} = [1/(1 + r)]p_t$, which is the same thing as

$$\beta = \frac{p_{t+1}}{p_t} = \frac{1}{1+r} \tag{3.6}$$

The change in the endowment from period t to period $t + 1$ obviously corresponds to any economic growth (or decline) occurring over that interval, so that

$$\frac{\omega_{t+1}}{\omega_t} = 1 + g \tag{3.7}$$

where g is the rate of growth of the economy. Given that the utility function (3.1) is of such a simple time-additive functional form, it is natural to interpret the utility weight a_2 simply as the weight a_1 reduced by a factor based on the subjective rate of time preference of the agent. This is consistent with Samuelson's original (1937) formulation of the DU model (Frederick et al. 2002). If this subjective rate of time preference is given by δ, then the relationship is

$$\frac{a_2}{a_1} = \frac{1}{1+\delta} \tag{3.8}$$

Returning to equation (3.5) and also using the approximation that $(1 + z)^\gamma \approx 1 + \gamma z$ for small z, we have

$$r = \delta + (1-b)g \tag{3.9}$$

Finally, for utility function (3.1), easy algebra shows that the intertemporal elasticity of substitution σ is given by $1/(1-b)$, and α, the elasticity of the marginal utility of consumption,[9] is given by $(1-b)$. This yields the familiar "descriptive" relationship for the market rate of interest

$$r = \delta + \alpha g \qquad (3.10)$$

provided the requisite approximations hold.[10]

Equation (3.10) (or 3.9) already carries within itself the seeds of a critique of current integrated assessment practice. Suppose that the b parameter lies somewhere between zero (corresponding to a Cobb–Douglas utility function) and $-\infty$ (corresponding to a Leontief utility function). Then $(1-b)$ is positive, and the market rate of interest will be greater or less than the rate of subjective time preference, depending on whether the economy is showing positive or negative growth. The ordinary practice in integrated assessment is to assume that the economy's growth will be positive due to technological progress and capital accumulation. However, it is possible under catastrophic climate scenarios that the amount of the "good" available for consumption could fall. A slowdown in the rate of growth of output brought about by environmental decline would lead to a lower interest rate than would prevail absent the environmental damage, and this possibility has been noted by Tol (1994) and Amano (1997). There is no basis for treating current market interest rates as being "descriptive" of future interest rates if the rate of growth of the economy, measured properly to include environment as part of the material standard of living, is subject to change over time.

In what follows, this theme will be explored in greater depth by specifying a second good, "environment," with a time path of its own that may be distinct from the time path of the endowment of the "ordinary" economic good. The next step, however, is to examine the case of more than one type of consumer. We shall see that this most basic generalization of the conventional framework explodes the prescriptions of existing climate policy models.

3.3.2 Two agents, one good, two time periods: allowing for intergenerational differences in perspective

To move beyond the "representative agent" in an intertemporal setting, we specify preferences for two distinct agents:

$$u_1 = a_{11}\frac{x_{1,t}^b}{b} + a_{12}\frac{x_{1,t+1}^b}{b} \qquad (3.11)$$

and

$$u_2 = a_{21} \frac{x_{2,t}^b}{b} + a_{22} \frac{x_{2,t+1}^b}{b} \tag{3.12}$$

The notation here is entirely analogous to that of the single-agent utility function (3.1); the initial subscript on the x variables indicates which of the two agents' consumption is being denoted. Both agents have the same substitutability parameter; this assumption is made purely for mathematical convenience. For our purpose here, it is sufficient to distinguish the agents by their utility weights a_{ij} and by their endowments of the good in the two periods.

Each agent's maximization of utility subject to the two-period budget constraint leads to four individual demand functions, one for each agent and each dated good. For example, the demand function for agent 1 for the good in period t is given by

$$x_{1,t}^* = \frac{\left(\frac{a_{12}p_t}{a_{11}p_{t+1}}\right)^{\frac{1}{b-1}}(p_t\omega_{1,t} + p_{t+1}\omega_{1,t+1})}{p_t\left(\frac{a_{12}p_t}{a_{11}p_{t+1}}\right)^{\frac{1}{b-1}} + p_{t+1}} \tag{3.13}$$

The endowments are indexed such that $\omega_{i,j}$ = the ith individual's endowment of the good in period j. The market-clearing equations take account of the fact that there are two agents demanding the good in each period. Thus, for example, the market-clearing requirement for period t is that

$$x_{1,t}^* + x_{2,t}^* - \omega_{1,t} - \omega_{2,t} = 0 \tag{3.14}$$

The same price normalization, $\beta = p_{t+1}/p_t$, will be used to find the equilibrium inflation ratio and market rate of interest.

The two-agent model differs significantly from the single-agent model in two respects. First, the equilibrium interest rate depends on both the characteristics of the individual agents' preferences and on the distribution of the good between agents, as well as on the intertemporal distribution of the good. Second, *more than one equilibrium is not only possible, but is quite likely given a realistic interpretation of the nature of the differences between the two agents' preferences.*

To demonstrate the first point as simply as possible, consider the case in which the agents' preferences are Cobb–Douglas, that is, where $b = 0$. In that case, the equilibrium inflation ratio β is given by

$$\beta^* = \frac{a_{12}(a_{21} + a_{22})\omega_{1,t} + a_{22}(a_{11} + a_{12})\omega_{2,t}}{a_{11}(a_{21} + a_{22})\omega_{1,t+1} + a_{21}(a_{11} + a_{12})\omega_{2,t+1}} \qquad (3.15)$$

The endowments in period t appear in the numerator of (3.15), while those of period $t + 1$ appear in the denominator. However, unless the utility weights are strictly proportional across agents (so that the utility functions are in effect identical), the inflation ratio (and hence the interest rate) *will depend on the distribution of the endowments between the two agents in each period.*

This characteristic of the equilibrium, which quite obviously carries over in the case of the more general CES (and other) utility functions, is highly significant for climate policy. The crucial endogenously determined interest rate cannot be known without knowing something about both the *preferences of the agents* and also the *distribution of wealth between them* in every period. Considerations of "efficiency" cannot be separated from question of "equity." Because the distribution of the endowment in each period is quite obviously policy-dependent, comparative static assessments of the alternative policies (such as a carbon tax, a permit auction system, or efficiency standards) cannot validly be undertaken without specification of the distributional consequences of the policies. For, as we have seen in Chapters 1 and 2, the distribution of endowments is quintessentially a policy question, involving as it does the assignment of "rights" of various kinds to members of the polity.

Now consider how this two-agent model might be utilized to represent a climate policy simulation in which long periods of time (in particular, periods spanning more than a single generation) might be interpreted. The key question here is whether the extended time period imposes any restrictions on the parameters of the preference functions and the endowments. Suppose we consider time periods long enough to separate the agents into two distinct groups, the "present" and the "future." The durations of these periods are sufficiently long that they encompass different generations. In this case, it makes sense to imagine that both the parameters and endowments are "mirror images" of each other: The "present" population cares mainly about itself, and is endowed mainly with "present" (time t) goods, so that $a_{11} \gg a_{12}$ and $\omega_{1,t} \gg \omega_{1,t+1}$. The opposite holds for the "future" population; $a_{21} \ll a_{22}$ and $\omega_{2,t} \ll \omega_{2,t+1}$. As a further simplification, suppose the preferences and endowments are completely symmetrical, so that $a_{11} = a_{22}$, $a_{12} = a_{21}$, $\omega_{1,t} = \omega_{2,t+1}$, and $\omega_{1,t+1} = \omega_{2,t}$. With these values, the equilibrium inflation ratio in the Cobb–Douglas case is 1 and the market rate of interest is 0.

Preferences and endowments of this type are precisely what one might expect under the useful fiction of "all transactions taking place at the initial moment of time." If we were to imagine all the people who will ever live gathering together in the imaginary marketplace to conduct their transactions at the beginning of time, the preferences and endowments of people who will be alive at different points in time have to be of this pattern – greater weight placed on consumption occurring when they are alive, and most of their "endowment" occurring during their own lifetimes. If the pattern is entirely symmetrical with Cobb–Douglas utilities, the interest rate is zero, reflecting the equal bargaining power (and similar "fundamental" situation) of the people living at different times.

Why do conventional integrated assessment models fail to reflect this? It is because all the agents in those models manifest a common "subjective rate of time preference." Such an assumption may be suitable for examining the interactions of people who are all alive today, but it is not suitable if we are to imagine people alive at very different times. To see how the conventional models can be misleading, consider what happens if both agents exhibit "conventional" preferences, with the utility weight on period $t + 1$ consumption equal to a discount factor times the weight on current consumption. That is, assume in this "present-oriented" case that

$$a_{12} = \rho a_{11} \qquad a_{22} = \rho a_{21} \qquad (3.16)$$

with $\rho < 1$. In this case, the structure of the two consumers' utility functions is identical. Both have the same subjective rate of time preference. The utility functions differ only by a multiplicative constant (a_{11}/a_{21}) and the agents are identical. This case collapses to that of a single agent, and the inflation ratio is given by (3.5). The conventional "Ramsey rule" holds and the interest rate depends only on the subjective rate of time preference, the growth rate of the economy, and the substitution parameter.

However, if the agents have different rates of time preference, say ρ_1 and ρ_2, in the Cobb–Douglas case ($b = 0$) the equilibrium inflation ratio is given by

$$\beta = \frac{\rho_1(1 + \rho_2)\omega_{1,t} + \rho_2(1 + \rho_1)\omega_{2,t}}{(1 + \rho_2)\omega_{1,t+1} + (1 + \rho_1)\omega_{2,t+1}} \qquad (3.17)$$

and once again the interest rate will depend on how the endowments are distributed across agents. Suppose agent 2 has a higher rate of time preference than agent 1 (so that $\rho_1 > \rho_2$). Then it can be seen from (3.17) that a redistribution from agent 1 to agent 2 in either period will increase the market rate of interest, and conversely in the case of a redistribution in favor of agent 1 in either period.

Now, consider more general CES utility functions. Let us focus on the case in which the utility functions and endowments are symmetrical. The (perhaps) surprising result here is that there will in general be multiple equilibria in this simple model economy. The reason is that the mathematical structure of the model is identical to that of the two-consumer, two-good model of Kehoe (1998) that was discussed in Chapter 2. The economy can exhibit a positive, negative, or zero market rate of interest if the intertemporal elasticity of substitution is sufficiently low. This makes perfect sense, given the fiction of all agents transacting at the beginning of time; the equilibrium can favor the "present" individuals with a positive market rate of interest, the "future" individuals with a negative market rate of interest, or be neutral between them (zero rate of interest). Prices and utilities will differ for the individuals in the different equilibria. Yet all three equilibria are Pareto-optimal perfectly competitive equilibria. The multiplicity is a consequence of the fact that *there is no preferred vantage point in time* if all the agents are imagined to transact simultaneously.

Conventional climate/economy models assume away this possibility of multiple equilibria by (1) specifying a "representative agent" (or agents with similar time vantage points) and (2) assuming that the intertemporal elasticity of substitution in consumption is relatively high (typically that the utility functions are Cobb–Douglas or logarithmic – see Table 2.4). However, nothing in the logic of general equilibrium requires these restrictions. Indeed, the more plausible interpretation of the meaning of the utility functions of the individuals who exist at different times is that these utility functions are "symmetric" (or, more exactly, that each agent's utility function is "centric" to that agent's time period) in both utility weights and endowments. This diversity of the utility functions and the time patterns of the endowments, combined with a lower degree of intertemporal substitutability than is typically assumed, is sufficient to produce multiple equilibria. The properties of these equilibria are, as Chapter 2 demonstrated, relatively unexplored in empirical work (because current empirical practice rules them out by assumption). Furthermore, economics offers little or no guidance as to which equilibrium or equilibria the social system will select. As

Mas-Colell et al. put it, this "is a manifestation of a serious shortcoming [in economic analysis] – the lack of a theory of equilibrium selection" (1995, p. 620).[11] It may be that current political institutions lead to a "present-centric" equilibrium, but that is not to say that such an equilibrium is an *economic* choice. Instead, it is a reflection of the way present-day political processes are structured to favor one group (those alive now) over all others (the future generations).

3.3.3 Two agents, two goods, two time periods

Needless to say, all of these results carry over to a situation in which there is more than one good as well as more than one type of agent. There are even more permutations of possible outcomes in this case. In addition to all the factors discussed in the previous section, the market rate of interest can depend also on the distribution of *endowments in the environmental good* across the agents, in either period. That is, policies that create or destroy rights to environmental services in the future can affect current relative prices of environmental and ordinary goods, the market rate of interest, and the current income distribution.

Consider again the minimal case of two agents, two time periods, and now two goods rather than one. The utility functions of agents 1 and 2 will be given by

$$u_1 = a_{11}\frac{x_{1,t}^b}{b} + a_{12}\frac{x_{1,t+1}^b}{b} + c_{11}\frac{y_{1,t}^b}{b} + c_{12}\frac{y_{1,t+1}^b}{b} \tag{3.18}$$

and

$$u_2 = a_{21}\frac{x_{2,t}^b}{b} + a_{22}\frac{x_{2,t+1}^b}{b} + c_{21}\frac{y_{2,t}^b}{b} + c_{22}\frac{y_{2,t+1}^b}{b} \tag{3.19}$$

where here x_{ij} represents the consumption of the first good ("ordinary economic goods and services") by agent i in period j, and y_{ij} represents the consumption of the second good "environmental goods and services" by agent i in period j. Again for simplicity, all of the substitution parameters are assumed to be equal.[12] The endowments of the first good are denoted by ω_{ij}, and of the second good by ε_{ij}.

In this setup, there will be four market-clearing equations (one for the excess demand for each good in each period) and four prices (one for each good in each period). Once again, a price normalization is required, so in addition to the familiar inflation ratio there will be two relative prices,

$$\beta = \frac{p_{t+1}}{p_t} \qquad \phi = \frac{q_t}{p_t} \qquad \psi = \frac{q_{t+1}}{p_{t+1}} \tag{3.20}$$

where p_j is the price of the first good in period j and q_j is the price of the second good in period j.

The CES utility versions of this model begin to exceed the capacity of symbolic equation solvers such as Mathematica, but the main features of the equilibria can be discerned from special cases and numerical examples. Thus, with Cobb–Douglas utility functions, the (unique) equilibrium for $\{\beta, \phi, \psi\}$ is given by

$$
\begin{aligned}
\psi = (\omega_{1,1+t}((c_{12}(a_{21} + a_{22} + c_{21} + c_{22})\varepsilon_{1,t} \\
+ (c_{11}c_{22} + c_{12}(a_{21} + a_{22} + c_{22}))\varepsilon_{2,t})\omega_{1,t} \\
+ ((a_{11}c_{22} + c_{12}(a_{22} + c_{21} + c_{22}))\varepsilon_{1,t} \\
+ (a_{22}c_{12} + (a_{11} + c_{11} + c_{12})c_{22})\varepsilon_{2,t})\omega_{2,t}) + (a_{21}c_{12}(\varepsilon_{1,t} + \varepsilon_{2,t})\omega_{1,t} \\
+ c_{12}(c_{21}\varepsilon_{1,t} + c_{22}(\varepsilon_{1,t} + \varepsilon_{2,t}))(\omega_{1,t} + \omega_{2,t}) + c_{22}(a_{11}(\varepsilon_{1,t} + \varepsilon_{2,t})\omega_{2,t} \\
+ c_{11}\varepsilon_{2,t}(\omega_{1,t} + \omega_{2,t}) + a_{12}(\varepsilon_{1,t} + \varepsilon_{2,t})(\omega_{1,t} + \omega_{2,t})))\omega_{2,1+t}) \\
/(a_{12}(a_{21}(\varepsilon_{1,t} + \varepsilon_{2,t})(\varepsilon_{1,1+t} + \varepsilon_{2,1+t})\omega_{1,t} \\
+ (c_{22}\varepsilon_{1,1+t}(\varepsilon_{1,t} + \varepsilon_{2,t}) + c_{21}\varepsilon_{1,t}(\varepsilon_{1,1+t} + \varepsilon_{2,1+t}) \\
+ a_{22}(\varepsilon_{1,t} + \varepsilon_{2,t})(\varepsilon_{1,1+t} + \varepsilon_{2,1+t}))(\omega_{1,t} + \omega_{2,t})) \\
+ a_{22}(c_{11}\varepsilon_{2,t}(\varepsilon_{1,1+t} + \varepsilon_{2,1+t})(\omega_{1,t} + \omega_{2,t}) \\
+ (\varepsilon_{1,t} + \varepsilon_{2,t})(a_{11}(\varepsilon_{1,1+t} + \varepsilon_{2,1+t})\omega_{2,t} + c_{12}\varepsilon_{2,1+t}(\omega_{1,t} + \omega_{2,t}))))
\end{aligned}
\tag{3.21}
$$

$$
\begin{aligned}
\phi = (\omega_{1,1+t}((c_{11}(a_{21} + a_{22} + c_{21} + c_{22})\varepsilon_{1,1+t} + ((a_{21} + a_{22})c_{11} \\
+ (c_{11} + c_{12})c_{21})\varepsilon_{2,1+t})\omega_{1,t} + ((a_{11}c_{21} + c_{11}(a_{22} + c_{21} + c_{22}))\varepsilon_{1,1+t} \\
+ (a_{22}c_{11} + (a_{11} + c_{11} + c_{12})c_{21}\varepsilon_{2,1+t})\omega_{2,t}) \\
+ (((a_{12}c_{21} + c_{11}(a_{21} + c_{21} + c_{22}))\varepsilon_{1,1+t} \\
+ (a_{21}c_{11} + (a_{12} + c_{11} + c_{12})c_{21})\varepsilon_{2,1+t})\omega_{1,t} \\
+ (((a_{11} + a_{12} + c_{11})c_{21} + c_{11}c_{22})\varepsilon_{1,1+t} \\
+ (a_{11} + a_{12} + c_{11} + c_{12})c_{21}\varepsilon_{2,1+t})\omega_{2,1+t}) \\
/(a_{11}(a_{22}(\varepsilon_{1,t} + \varepsilon_{2,t})(\varepsilon_{1,1+t} + \varepsilon_{2,1+t})\omega_{1,1+t} \\
+ a_{21}(\varepsilon_{1,t} + \varepsilon_{2,t})(\varepsilon_{1,1+t} + \varepsilon_{2,1+t})(\omega_{1,1+t} + \omega_{2,1+t}) \\
+ (c_{22}\varepsilon_{1,1+t}(\varepsilon_{1,t} + \varepsilon_{2,t}) + c_{21}\varepsilon_{1,t}(\varepsilon_{1,1+t} + \varepsilon_{2,1+t}))(\omega_{1,1+t} + \omega_{2,1+t})) \\
+ a_{21}(c_{11}\varepsilon_{2,t}(\varepsilon_{1,1+t} + \varepsilon_{2,1+t})(\omega_{1,1+t} + \omega_{2,1+t}) \\
+ (\varepsilon_{1,t} + \varepsilon_{2,t})(a_{12}(\varepsilon_{1,1+t} + \varepsilon_{2,1+t})\omega_{2,1+t} + c_{12}\varepsilon_{2,1+t}(\omega_{1,1+t} + \omega_{2,1+t}))))
\end{aligned}
\tag{3.22}
$$

$$\beta = (a_{12}(a_{21}(\varepsilon_{1,t} + \varepsilon_{2,t})(\varepsilon_{1,1+t} + \varepsilon_{2,1+t})\omega_{1,t} + (c_{22}\varepsilon_{1,1+t}(\varepsilon_{1,t} + \varepsilon_{2,t})$$
$$+ c_{21}\varepsilon_{1,t}(\varepsilon_{1,1+t} + \varepsilon_{2,1+t}) + a_{22}(\varepsilon_{1,t} + \varepsilon_{2,t})(\varepsilon_{1,1+t} + \varepsilon_{2,1+t}))(\omega_{1,t} + \omega_{2,t}))$$
$$+ a_{22}(c_{11}\varepsilon_{2,t}(\varepsilon_{1,1+t} + \varepsilon_{2,1+t})(\omega_{1,t} + \omega_{2,t})$$
$$+ (\varepsilon_{1,t} + \varepsilon_{2,t})(a_{11}(\varepsilon_{1,1+t} + \varepsilon_{2,1+t})\omega_{2,t} + c_{12}\varepsilon_{2,1+t}(\omega_{1,t} + \omega_{2,t}))))$$
$$/(a_{11}(a_{22}(\varepsilon_{1,t} + \varepsilon_{2,t})(\varepsilon_{1,1+t} + \varepsilon_{2,1+t})\omega_{1,1+t}$$
$$+ a_{21}(\varepsilon_{1,t} + \varepsilon_{2,t})(\varepsilon_{1,1+t} + \varepsilon_{2,1+t})(\omega_{1,1+t} + \omega_{2,1+t})$$
$$+ (c_{22}\varepsilon_{1,1+t}(\varepsilon_{1,t} + \varepsilon_{2,t}) + c_{21}\varepsilon_{1,t}(\varepsilon_{1,1+t} + \varepsilon_{2,1+t}))(\omega_{1,1+t} + \omega_{2,1+t}))$$
$$+ a_{21}(c_{11}\varepsilon_{2,t}(\varepsilon_{1,1+t} + \varepsilon_{2,1+t})(\omega_{1,1+t} + \omega_{2,1+t})$$
$$+ (\varepsilon_{1,t} + \varepsilon_{2,t})(a_{12}(\varepsilon_{1,1+t} + \varepsilon_{2,1+t})\omega_{2,1+t} + c_{12}\varepsilon_{2,1+t}(\omega_{1,1+t} + \omega_{2,1+t})))) \tag{3.23}$$

It is clear that the equilibrium depends on all the parameters – the utility weights of each agent and the endowments of each good to each agent in each period.

Various other specialized cases (symmetrical utility, conventional utility weights with "forward-looking" time preferences, etc.) yield results similar to what was obtained in section 3.3.2 – period-centric utility functions and endowments plus low intertemporal substitutability can produce multiple equilibria, for example. There are now many more parameters that can be varied independently, however, and the curse of dimensionality rears its head so that a comprehensive numerical exploration of the possibilities is not feasible. (In the 2-agent, 1-good case there were four utility weights, four endowments, and one substitution parameter or nine in all; in the 2-agent, 2-good, 2-period case there are eight utility weights, eight endowments, and one substitutability parameter or 17 parameters.[13])

Several key points emerge from the discussion so far:

- The prices (including the market rate of interest) and corresponding utilities (welfare measure) and incomes depend intrinsically on the endowments. These endowments, in turn, are policy-determined. The State assigns and enforces a particular set of property rights. In circumstances in which hitherto undefined property rights are increasing in importance, it is analytically incorrect and politically misleading simply to treat the property rights that have been historically defined as the only ones that matter; the "new" property rights (including assignment of climate rights) should be an intrinsic component of the analysis.
- Multiple equilibria are possible, when models of intertemporal general equilibrium are interpreted properly as embodying the

preferences of individuals who exist at different points in time. In particular, there is no single preferred time vantage point, such as the present. Current integrated assessment practice rules out the possibility of different intertemporal vantage points by assumption, and thereby produces analytical results that are biased in favor of the present population.

- The so-called "descriptive" characterization of the market rate of interest is in reality a highly restrictive special case. In particular, it is inconsistent with people living at different times having different utility functions, and it fails to reveal how the market rate of interest depends on distributions of endowments of the various kinds of goods in every period.

3.4 Overlapping generations: transactions occur in the present, accounting for past history and expectations

3.4.1 A two-period OLG model with one agent in each generation

In keeping with the approach taken in the previous sections, our discussion of OLG models will begin with an examination of the simplest case. In this model, there is one agent in each generation. The agents live for two time periods. There is a single good, and each agent is endowed with ω_0 of the good during the first period of his life and ω_1 of the good during the second period of his life. The agent born in period t consumes $x_{t,t}$ in period t, and $x_{t,t+1}$ in period $t + 1$. In this notation, the first subscript denotes the period in which the agent is born, while the second subscript denotes the period in which the consumption takes place. The economy can thus be described by Table 3.1, showing the consumption of the good by different agents at different points in time. In this table, total consumption of the good in each time period is given by column sums; thus, consumption in period t is the sum of the consumption by the "young" (those born in t) in period t, which is $x_{t,t}$, and the consumption of the "old" (those born in $t - 1$) in period t, which is $x_{t-1,t}$. The boldface entries within Table 3.1 show a complete account of the consumption values for time periods $t - 1$ through $t + 2$; the ellipses outside indicate that the table could be extended backward or forward in time by filling in the appropriate spaces.

The utility of an agent is given by a constant elasticity of substitution utility function depending on the agent's consumption in each time period, e.g.,

Table 3.1   Time structure of consumption in OLG model

Period born	Time period					
	$t-2$	$t-1$	t	$t+1$	$t+2$	$\ldots$
$t-3$	$\ldots$					
$t-2$	$x_{t-2,t-2}$	$x_{t-2,t-1}$				
$t-1$		$x_{t-1,t-1}$	$x_{t-1,t}$			
t			$x_{t,t}$	$x_{t,t+1}$		
$t+1$				$x_{t+1,t+1}$	$x_{t+1,t+2}$	
$t+2$					$x_{t+2,t+2}$	$x_{t+2,t+3}$
$t+3$						$\ldots$

$$u(x_{t,t}, x_{t,t+1}) = \frac{a_1 x_{t,t}^b}{b} + \frac{a_2 x_{t,t+1}^b}{b} \tag{3.24}$$

Utility depends on consumption in the two periods during which the agent is alive, and a_1, a_2, and b are the parameters of the utility function. For the agent born in t, the price of the good in period t is given by p_t, and the price *expected* in the next period is p_{t+1}^e. For the agent born in $t-1$, the market price that prevailed in $t-1$ is known and the period-t maximization problem of this agent does not depend on expectations. This is the first essential difference between the OLG model and the "simultaneous transactions" model of the previous section. In the OLG framework, the market clears in each period, given past history (that is, the history of past prices p_{t-1}, p_{t-2}, ...) and the agents' expectations of what the future periods' market-clearing prices will be (for those agents who will be alive in the future period or periods). This modeling twist involves a gain and a loss. The gain is that the model is more realistic than the simultaneous transactions model, because it confines the equilibration to each successive period and does not require that all periods' markets clear at the beginning of time. The loss is that the equilibrium now must depend on expected future prices, because the final equilibrium conditions in the future periods cannot be known in advance by the agents. We shall see that this crucial difference leads to an additional (and irreducible) source of multiplicity in OLG models.

Each agent must make plans based on his lifetime endowment. Thus, the budget constraint for the agent born in t is given by

$$p_t x_{t,t} + p_{t+1}^e x_{t,t+1} \leq p_t \omega_0 + p_{t+1}^e \omega_1 \tag{3.25}$$

where ω_0 is the endowment of the agent in period t (when he is "young"), and ω_1 is the endowment of the agent in period $t + 1$ (when he is "old"). For the agent born in $t - 1$, the budget constraint is

$$p_{t-1}x_{t-1,t-1} + p_t x_{t-1,t} \leq p_{t-1}\omega_0 + p_t\omega_1 \tag{3.26}$$

because we are considering the simple case in which the endowments when "young" and "old" are unchanged for each generation.[14]

Maximization of the utility function (3.24) subject to their budget constraints by the two agents who are alive during period t yields their respective demands for the good in period t. Finally, we specify that the good is perishable, so that it cannot be carried over from one period to the next. Under these circumstances, the market equilibrium in each period is determined by the condition that the sum of agents' demands must equal the total amount of the good available. Thus, market equilibrium in period t requires that

$$x_{t,t}^* + x_{t-1,t}^* = \omega_0 + \omega_1 \tag{3.27}$$

where the starred values are the agents' demand functions obtained from their budget-constrained utility maximization. Note here that the amount of the good available in period t is the sum of the endowments of the good possessed by the "young" agent born in t and the "old" agent born in $t - 1$.

As usual, a price normalization rule must be chosen, and the natural one is to normalize in terms of the inflation ratios, $\beta_{t+1}^e = p_{t+1}^e/p_t$ and $\beta_t = p_t/p_{t-1}$. The equilibrium in period t is then defined as the value of β_t that solves the market-clearing equation

$$\frac{\omega_0 + \omega_1\beta_t}{\left(\frac{a_2}{a_1\beta_t}\right)^{\frac{1}{b-1}} + \beta_t} + \frac{\left(\frac{a_2}{a_1\beta_{t+1}^e}\right)^{\frac{1}{b-1}}(\omega_0 + \omega_1\beta_{t+1}^e)}{\left(\frac{a_2}{a_1\beta_{t+1}^e}\right)^{\frac{1}{b-1}} + \beta_{t+1}^e} - \omega_0 - \omega_1 = 0 \tag{3.28}$$

for some particular expectation-formation mechanism. Before considering the effect of different expectations mechanisms, consider first the case of the *rational expectations steady state*. A steady-state equilibrium is one in which the inflation ratio remains constant, and is perfectly anticipated by the agents. Thus, the steady state embodies a form of "rational expectations," because the expected values of β coincide in every period with the realized values. Equation (3.28) has two steady-state solutions, given by

$$\beta_1^* = 1 \tag{3.29}$$

and

$$\beta_2^* = \left(\frac{a_2}{a_1}\right)\left(\frac{\omega_0}{\omega_1}\right)^{1-b} \tag{3.30}$$

The first point to be noted is that the steady-state equilibrium of this economy is not unique. Unlike in the static case of Chapter 2, this nonuniqueness is not a function of the particular values of the parameters; there are two steady states for *any* valid parameter set. The second observation is that the steady states have different utility values. In particular, the steady state β_1^*, usually referred to as the "Golden Rule" steady state, has the higher utility value. It is also clear that the interest rate in the β_2^* steady state depends on the pattern of the endowments over time. This feature of OLG models has been noted before (Howarth and Norgaard 1992, Gerlagh and van der Zwaan 2001a, b). This dependence of the market rate of interest on the pattern of endowments (or production) is intrinsic to intertemporal general equilibrium analysis.

The steady state β_2^* is referred to as the "autarkic" steady state because it has the property that each agent consumes exactly his own endowment in each period. Why does the "Golden Rule" steady state have higher utility? In it, the agents exchange some of their endowments in t so that each can have a higher utility value. This obviously can be beneficial for the "old" agent (assuming that the endowment is larger during the "young" segment of life), but what about the "young" agent? If the young agent exchanges some of his endowment with the old agent, the young agent's consumption in t is obviously less than it would have been had he chosen to consume his entire endowment. He achieves a higher level of utility because in the *next* period, he will be old and will benefit from the same kind of exchange, consuming more than his endowment when old. But what guarantee is there that in $t + 1$ the young agent born in $t + 1$ will agree to this deal? The young agent in t can only hope that the same kind of exchange can be arranged by the agents born in $t + 1$, $t + 2$, etc. One kind of institution that might enable this to occur is some form of Social Security, in which each agent can treat the utility-increasing exchange as an entitlement. Another way to accomplish the ongoing social contract is through the existence of fiat money, whose value is "guaranteed" by continuing agreement across the generations. Both of these mechanisms were discussed in Samuelson's original (1958) model laying out the OLG framework.

Before going on to treat the important question of expectations formation and the path dependence of equilibria in OLG models, let us first examine generalizations of the model to include more than one type of agent per generation.

3.4.2 Generalizations: OLG models with more than one type of agent and more than two periods

How do these results change if each generation contains more than one kind of agent? We shall see momentarily that, just as in the Arrow–Debreu models, the equilibria will depend on the distribution of endowments across the different agents. Also, multiple equilibria can arise, depending on whether these agents are both "present-oriented" or not. This distinction will be seen to lead naturally to models in which there are more than two periods.

First consider a two-period OLG model with two different agents in each generation. By now, the utility functions, budget constraints, price normalizations, and market-clearing conditions will be familiar. The utility functions are

$$u_{1,t} = a_{11}\frac{x_{1,t}^b}{b} + a_{12}\frac{x_{1,t+1}^b}{b} \tag{3.31}$$

$$u_{2,t} = a_{21}\frac{x_{2,t}^b}{b} + a_{22}\frac{x_{2,t+1}^b}{b} \tag{3.32}$$

These utility functions pertain to the individuals born in t; their utility depends on their consumption in the two periods of their lives. Similar utility functions can be written for individuals born in any other period. The equilibrium in period t is again determined by the market-clearing equation specifying that the sum of the demands of all the agents alive in t equals the total amount of the good available. The equilibrium price ratio β_t will again depend on the "fundamentals" (preferences and endowments) and on the mechanism by which the agents form expectations about the future. The steady-state rational expectations equilibrium is given by the value or values of β that remain constant over time.

Now it may or may not be surprising that the mathematical form of the equation for the steady-state equilibrium in this case is the same as in the Arrow–Debreu case with two agents, one good, and two time periods. The utility functions (3.31) and (3.32) have the same form as those of (3.11) and (3.12); there are four endowments that make up the supply of the good in period t: those possessed by each of the two young agents born in t and those possessed by each of the two old agents born in $t - 1$. The same price normalization can be used, with its same interpretation as a transformation of the market rate of interest. The same findings as in the Arrow–Debreu case carry over: the interest rate will in general depend on the distribution of endowments across agents in each time period, and multiple equilibria (in addition to the Golden Rule steady state) are possible if the two agents have "symmetrical" preferences and the intertemporal elasticity of substitution is not too high.

What does it mean in this case for the utility functions of the two agents to be "symmetrical"? In the Arrow–Debreu case, the interpretation was that agents living during different time periods would normally weight most heavily the consumption they enjoy when they are alive, and that their endowments would accrue to them mainly during their respective lifetimes. In the OLG framework, we can imagine people whose lifetimes overlap, with each period of the model corresponding to a particular span of the overlapping years. Now it is possible in this setting to think of two kinds of people: "present-oriented" people whose utility is grounded in short-term pleasures and whose most important endowment is in the near term. Imagine a hedonist living "for the moment" whose most important asset is physical attractiveness. Contrast this with a "future-oriented" person, someone who looks forward to a high income and high productivity later in life, based on the accumulation of human capital and job experience. These two kinds of individuals could be thought of as having the type of symmetrical preferences that lead to multiple equilibria.

It is possible to carry the thought experiment farther, however. In reality, the periods of overlap of the generations are not sharp and discrete as in the model. People are being born and are dying continuously. In any given interval of years, there is a distribution of people of various ages. If we think of the "young" and the "old" people in any such interval (corresponding to the two "generations" of the model who are alive at any time t), we realize that within the "young" there will be people with ages ranging from newborn to middle-aged. Similarly, within the "old" there will be people of ages ranging from middle-aged to octogenarian (and older). It is plausible for the "young" group then to be made up of diverse agents with different weights on their consumption (and different endowments) when "young" and when "old." A middle-aged person who just happened to fall into the "young" cohort would have a low endowment when young and would place relatively low weight on his consumption when young, compared to his utility weight and endowment when old. Conversely, a very young member of the "young" group would have the opposite kind of utility function (and possibly also endowment pattern). The point is that diversity within the generations (as captured crudely by the two-agent generalization) is to be expected.

An alternative way of modeling this type of age-specific diversity is to increase the number of periods in the individuals' lifetimes, and to allow for different kinds of life-cycle endowment patterns. This approach makes it possible to go back to the single-agent model. It has been explored in detail by Kehoe and Levine (1990) in an important paper dealing with the application of OLG models to problems of public policy in general. When the number of periods in the lifetime is three, Kehoe and Levine show that a single-good model calibrated roughly to realistic values for the life-cycle endowments, intertemporal elasticity of substitution, and subjective rate of time discount exhibits the same kind of multiple steady states (with corresponding different interest rate values) that we have seen in the two-agent models with symmetric utility functions. Kehoe and Levine specify three 20-year periods in the lifetime of each generation, a subjective rate of time preference of 3.5 percent per year, an intertemporal elasticity of substitution of 0.25,[15] and a not-implausible endowment pattern of $\{\omega_0, \omega_1, \omega_2\} = \{3, 12, 1\}$.[16] This model has three steady states in addition to the Golden Rule (with $r = 0$). The (annual) market interest rates in these three steady states are 9.1, 1.2, and –17.3 percent. As if this kind of indeterminacy were not daunting enough, we shall see in the next section that if an OLG economy approaches one of its steady states, the one it converges to depends both on the "initial conditions" (the state of expectations at the outset) and the expectations-formation mechanism of the agents.[17] Neither of these considerations is a "fundamental" of the economy.

3.4.3 The indispensable role of expectations

So far, only the steady-state rational expectations equilibria of the OLG models have been examined. Of course, restricting equilibria in this way is quite severe. The actual economy is so complicated that exact knowledge of its rational expectations equilibria is impossible in practice (Spear 1989, DeCanio 1999). What happens when the requirement that the economy be in its steady-state rational expectations equilibrium is relaxed? In the OLG model, given the history of the economy up to period t, it is possible to calculate the market-clearing equilibrium price ratio in period t given a mechanism for the formation of expectations about future prices.

To illustrate what can happen when this is done, the case of a 2-agent, 2-good, 2-period OLG model will be considered. We have already seen that this model can have multiple steady states if the intertemporal substitutability is low and the agents' utilities are "symmetric." Now let expectation formation take the familiar adaptive form (Nerlove 1958):

$$\beta_{t+1}^e = \beta_t^e + \lambda(\beta_t - \beta_t^e) \tag{3.33}$$

Equation (3.33) applies to the intertemporal price ratio variable β; a similar equation governs the expectations-formation mechanism of the relative price ratio ϕ.

In order to solve for the time path of the price variables, expectations have to be given initial values β_0^e and ϕ_0^e. Nothing about the fundamentals of the economy determines what these initial expectations values might be. Similarly, it is possible to imagine various values for the speed of adjustment parameter λ in equation (3.33) and its counterpart in the equation for ϕ_t^e. Numerical simulations show that many different possible starting points and speeds of adjustment are able to produce equilibrium paths that converge to the steady states. This is an illustration of the general proposition that there is a robust continuum of equilibria in OLG models (Geanakoplos 1989b).

This continuum of alternative equilibrium paths will exist even if there is only the one steady state other than the Golden Rule. Figure 3.1 shows three such paths for β and ϕ for different initial conditions for the $2 \times 2 \times 2$ model with Cobb–Douglas utility. Each path converges to the non-Golden Rule steady state, and different initial conditions would exhibit different transition paths.[18]

If there is more than one steady state, the path taken by the economy may converge to a steady state – depending on the starting point and speed of adjustment – or it may exhibit chaotic dynamics, depending on how the economic system "picks" the market equilibrium in each successive period. An example of the kind of behavior that is possible is given in Figure 3.2, for CES utility functions with a low degree of intertemporal and inter-good substitutability.[19] The economy whose path is depicted in Figure 3.2 has three steady states, corresponding to β values of 0.115, 1, and 8.686. (The corresponding annual interest rates would be 11.4, 0, and –10.2 percent if the duration of each time period is 20 years.) In each time period, the economy uses a numerical search method[20] to find the equilibrium, given the past price history and expectations formed according to (3.33). With this particular configuration of values, we see that the economy can spend long stretches of time (the 12 periods or 240 years beginning at $t = 17$ and the 5 periods or 100 years ending at $t = 37$, for example) near its "high interest rate" steady state. From time to time, however, the economy can jump to a

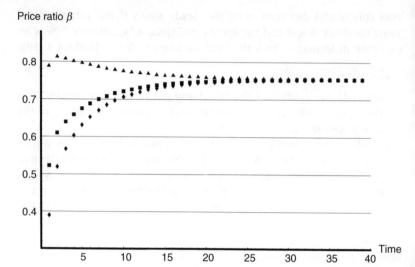

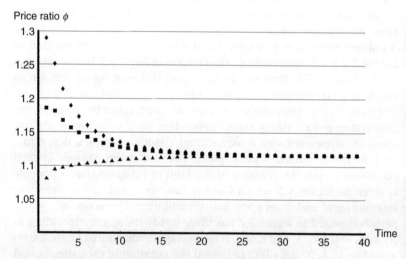

Figure 3.1   Alternative paths to the steady state: Cobb–Douglas case

"low interest rate equilibrium." At these times the market interest rate is in the vicinity of 1 percent per annum. Note that the "low interest rate equilibria" do *not* correspond to the future-oriented steady state (which has a negative interest rate). The interest rate is low but still positive.

Price ratio β

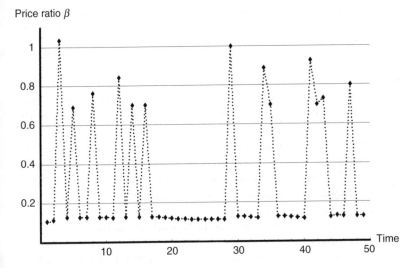

Figure 3.2   Chaotic equilibria: CES utility case

None of this should be taken to suggest that the probabilistic equilibrium choice mechanism underlying Figure 3.2 corresponds to what actually takes place in the real-world economy. But that is the point: *we do not know by what process the economy seeks equilibrium,* or moves from one equilibrium to another, or cycles between equilibria. If the economy had spent a long time close to one of its steady states (the "high interest rate" one, say) it is not even clear whether the agents could be aware of the existence of other possibilities. It might be that only a shock of some magnitude could bump the economy to an alternative equilibrium. If empirical economics had strong models of dynamics, it might be possible to specify how the "equilibrium choice" is made. But those dynamic models either do not exist or exist in only rudimentary form. But this example and the instability examples of Chapter 2 demonstrate that *dynamics matters fundamentally.* There is simply no scientific justification for assuming that the economy gravitates quickly to a unique and stable equilibrium.

All kinds of variations on these themes are possible. The most obvious direction in which extension might be made is to the three-period OLG model of the type employed by Kehoe and Levine. Given the role of wealth effects in creating the multiple equilibria, it is also possible to imagine models in which economic growth leads to ever-increasing endowments during the "productive" years. This might trigger a switch from a unique equilibrium to a situation of multiple equilibria. However, these possibilities should not distract us from the fundamental points:

- OLG models, like Arrow–Debreu models, can exhibit multiple equilibria. However, the multiplicities of the OLG models are even more pervasive. Multiple steady states are possible, corresponding to the case of multiple equilibria in the "full information" models in which all transactions take place at the beginning of time. In addition, the fact that the OLG models achieve equilibrium period by period means that different initial conditions and expectations adjustment mechanisms can lead to different transition paths to the steady states.

- Besides the fact that the equilibria depend on the parameters of the agents' utility functions and the patterns of their endowments, the equilibrium paths in OLG models also depend on how expectations are formed and modified. These expectation-formation mechanisms are distinct from the "fundamentals" of the economy. The "rational expectations" requirement is far too demanding of information and computational capacity to correspond to actual economic processes.

- If the economy does have multiple steady states, the expectations-formation mechanism can lead to chaotic equilibria, in which the prices and interest rate fluctuate according to each period's selection from among the multiple equilibria available to the agents in that period. This can lead to long stretches of time in the vicinity of distinct interest rates, so that the agents in the economy might not even be aware of the potential for an alternative equilibrium based on the "fundamentals" of the economy.

3.5 Models with a social planner or infinitely lived agents

Models in which the agents optimize over an infinite horizon, whether based on the interactions of a collection of infinitely lived decision-makers or on the choices made by a social planner having an infinite horizon, necessarily come up against the problem of how to formulate the optimization problem into a meaningful mathematical calculation. Maximization of infinite quantities has no meaning. In order to convert the consumption streams into finite quantities, some kind of constraint or weighting must be used. Thus Ramsey, the first economist to do rigorous work on this problem, required that there be an upper bound on the utility of consumers (which he denoted "bliss") so that there was no gain from increasing consumption beyond that point (Ramsey 1928). More typically, the infinite problem is converted to a finite one by a weighting scheme that ensures that the utility attached to future consumption diminishes geometrically as time goes on, so that the sum (or integral) of the entire infinitely long consumption stream is a finite quantity.[21]

This *mathematical* necessity has the strongest kinds of *substantive economic* implications, implications that are so unrealistic as to undermine the credibility of conclusions drawn from these assumptions. Consider what is involved in converting the infinite consumption stream of an agent into a finite quantity to be maximized. If we assume that the utility of the agent (or of the social planner, if the consumption stream is that of the entire community) is time-additively separable, then the total utility of the agent is given by

$$U = \sum_{t=0}^{\infty} \psi_t u_t(c_t) \tag{3.34}$$

where c_t is consumption in period t, u_t is the (possibly time-specific) utility of consumption in period t, and the ψ_t are the weights that are required to convert the otherwise infinite sum into a finite quantity. Ordinarily, the additional assumption is made that the utility function is the same for each time period (so that all the u_t are identical functions). One "natural" set of weights to convert the sum to a finite quantity are "discount factors" such that $\psi_t = \rho^t$ with $\rho < 1$. With these weights the value of U will be finite, so long as $u(c_t)$ does not grow too fast.

To put this into the form of an economic problem, there must be some constraint or limitation on the sequence of the c_t. In exchange models, this constraint is given in terms of the sequence of endowments ω_t and takes the form that total consumption cannot exceed the total endowment over the time path. In models with saving and production, the constraint takes the form that consumption depends on the amount of capital available to the economy, capital which can be accumulated only by diverting the economy's output to saving and away from current consumption. In either case, it is possible to specify conditions that enable the maximization problem to be well-defined and soluble.

This formulation is, however, deeply unsatisfactory. It has already been pointed out that no individual lives forever, so the "agent" engaged in the maximization cannot correspond to a real person. If the utility is supposed to correspond to some kind of social welfare function being maximized by an all-foreseeing benevolent social planner, we run up against the fact that in general it is not possible to construct a social welfare function. The infinite-horizon formulation of the problem requires specification of *cardinal utility* on the part of the agents in order to be able to combine those utilities to obtain a quantity that can be maximized. The reason is that the utility in equation (3.34) is summed over the entire (infinite) horizon of the model. In order for this adding up to be possible, the utilities at different points in time have to be comparable in the cardinal utility sense. In other words, in order to treat the succession of individuals living at different times as a single infinitely lived agent (or as a single social welfare function that the planner seeks to maximize), the utilities of the individuals have to be measured on a cardinal scale.

This notion of cardinal utility that allows utilities of different individuals to be added together or compared directly originated with the classical utilitarianism of Jeremy Bentham and John Stuart Mill, which sought to ground social ethics in those actions that bring about "the greatest happiness" (Brown 2002). However, such a concept is meaningless if the utilities of different people cannot be compared quantitatively. It is just this kind of "interpersonal comparison of utilities" that modern economics eschews. Direct interpersonal comparisons of utility have been rejected by economists since the 1930s, when it was realized that the theory of demand did not require cardinal utility.[22]

The invariance of observable demand behavior to monotonic transformations of the utility function enables modern welfare economics to adopt a strictly individualist stance.[23] The theory has testable implications even though it is not necessary to specify the exact form of the utility function. Certain features of the individual's preferences, such as the elasticity of substitution between goods, are intrinsic and invariant under any monotonic transformation of the utility function, and can in principle be recovered from data on individual behavior. What is not allowed is the aggregation of individual preferences into a single "social welfare function" of a particular mathematical form. Hence, welfare comparisons are permitted only according to the Pareto criterion, that one situation is better than another if and only if every individual is at least as well off in the first situation as in the second, and at least one person in the first situation is better off. Neoclassical welfare economics is predicated on the notion that individuals are unique and incommensurable; they interact through markets according to their own subjective interests, but their individual preferences cannot in any mathematically simple way be aggregated to form a social welfare function.

Now, the nonutilitarian philosophical purity of economics is not quite as clear-cut as it might seem from the discussion so far. Modern welfare economics also utilizes the "Kaldor–Hicks compensation principle," in which moving from one situation to another is considered to be welfare-improving if it is possible for the "winners" to compensate the "losers" with something left over. It should be noted that this is the working presumption of almost all contemporary policy analysis based on GDP or some measure of aggregate income – a situation with higher GDP is considered to be superior to one with less because in the new situation everyone could have the income he or she had enjoyed before, plus some people would have more. It is a nontrivial detail (to which we will return) that, as Amartya Sen (1979) has pointed out, this compensation principle may not count for much because in practice *the compensating transfers need not actually be made*. Policies can be implemented that have distributional consequences without social or political mechanisms to make the losers whole.

The individualist orientation of neoclassical economics also involves denial of the possibility that the well-being of others might affect one's own utility. This assumption, too, runs counter to both the empirical and subjective reality. (We care about other people, perhaps in proportion to their proximity to us.) There is also a substantial body of evidence suggesting that people care about their *relative* position in the wealth distribution. The presence of relative income in the utility function, or utility's dependence on positional goods, gives rise to externalities and social inefficiencies in a laissez-faire market system (Howarth 1996, 2000; see also section 2.2.2).

For our purposes, however, the main point is that cardinal utility is necessary for solution of the maximization problem of the infinitely lived agent. For example, one of the common versions of equation (3.34) as it appears in integrated assessment models has a social planner maximizing welfare over the infinite horizon by allocating consumption and investment over time to maximize

$$W = \int_0^\infty U[c(t)]e^{-\delta t}dt \tag{3.35}$$

where U is the utility function, $c(t)$ is the rate of consumption, and δ is the subjective discount rate. The problem has a well-defined solution if an appropriate production technology is specified that allows conversion of present consumption into future consumption, and if a boundary condition is met that keeps the value of W finite. The maximization problem becomes one of deciding on the division of output between savings and consumption at each point in time. Solution of the maximization problem under standard conditions yields the "Ramsey Rule" that was derived in the two-period model above (equation (3.10)). However, the solution is not invariant under monotonic transformations of U. Consider

$$W' = \int_0^\infty g\{U[c(t)]\}e^{-\delta t}\,\mathrm{d}t \tag{3.36}$$

where g is a positive monotonic function. It is easy to imagine g functions that would preclude a solution to (3.36) even if (3.35) could be solved. A particularly simple example is $g(U) = Ue^{\mu t}$, where $\mu > \delta$. Then W' is infinite so long as $U[c(t)]$ is increasing in t. In other words, if a Ramsey-type problem such as maximization of W in (3.35) is to be solved, the utility function must be cardinal.[24]

So far, we have been considering only an economy with a single good, "consumption." If there are multiple goods and multiple infinitely lived agents, the possibility of multiple solutions to the equilibrium maximization problem arises just as in the single time period case of Chapter 2 and in the finite number of periods case of section 3.3.3. Farmer (1999) has shown how this can be demonstrated in an exchange economy populated by a finite number of infinitely lived agents, with utility functions that are additively time-separable. The goods occurring in each time period can be thought of as different goods, so that this economy has a finite number of agents and an infinite number of goods. Negishi (1960) proved that this problem can be converted to one of solving a finite number of equations and unknowns. Furthermore, the reduced problem has the same mathematical structure as a system of market excess demand functions in a finite single-period economy. In general, the system will have multiple solutions.

Thus, even in the simplest type of infinite-horizon model – one with infinitely lived agents, time-separable utility functions, a discount factor less than one, and bounded wealth of the agents – multiple equilibria can exist. Even strong restrictions on the agents' utility functions and wealth endowments are not sufficient to guarantee uniqueness. The same problems appear as we have seen before, with the same absence of definitive empirical information that could point the way to restrictions sufficient to produce a unique solution. The unsatisfactory conceptualizations that are required for imagining infinitely lived agents or a benevolent social planner with an infinite horizon do not, in the end, rescue the model from the challenge posed by multiple equilibria.

3.6 Conclusions and implications for policy

Here are the main conclusions:

- In either the Arrow–Debreu or OLG framework, the equilibrium prices and allocations will depend in an essential way on the distribution of endowments of the different members of society. This conclusion is inescapable unless the diversity of agents is assumed out of existence by positing a "representative agent."
- In either the Arrow–Debreu or the OLG framework, a combination of plausible differences in the endowments and preferences of the different agents, in conjunction with empirically supported values of the intertemporal elasticity of substitution, will produce multiple equilibria, each of which is the possible outcome of a competitive market process. The means by which society settles on one such equilibrium out of the many possible ones is beyond the current understanding of economics.
- In the case of OLG models, there will in general be a continuum of optimal equilibrium paths, depending on the initial state of expectations and on the mechanism by which expectations are adjusted.

It should be noted that the first and third of these propositions hold regardless of whether the general equilibrium solution of the model is unique. There are other, and separate, reasons for preferring either the Arrow–Debreu or the OLG framework to that of the "social planner" models that assume an infinite horizon. Basically, these boil down to the point that the social planner approach requires interpersonal comparisons of utility (or, what amounts to the same thing, the assumption of cardinal utility) and that the social planner model requires an institutional framework that does not and cannot exist. The Arrow–Debreu model also rests on the fiction that all agents are able to transact at the beginning of time, but at least it does not impose a "dictatorship of the present" on the modeling outcome that is the result of applying the social planning approach.

A common thread in all the discussion so far is the crucial dependence of the second of the two results on the value of the intertemporal elasticity of substitution, $1/(1 - b)$. If this value is small and with plausible symmetric utility functions and endowments for the diverse members of society (or different generations), then multiple equilibria, with all that they entail for the impossibility of conventional comparative statics policy analysis, will be the rule. On the other hand, if the intertemporal elasticity of substitution is high, then equilibrium is likely to be unique whatever the pattern of utilities and endowments. The allocations, prices, and welfare of the different members of society will still depend on the policy-driven allocation of endowments (the endowment pattern being taken as a proxy for the various policy options available, and for the consequences of climate change), but there would be a better prospect that conventional integrated assessment models would yield reliable insight into the consequences of alternative policies.

There is a very large empirical literature on the value of the intertemporal elasticity of substitution (usually denoted by σ). As shown here and elsewhere, the value of the "social discount rate" or market rate of interest, so crucial to the evaluation of climate change damages and to the costs of mitigation policies, depends on the intertemporal elasticity of substitution[25] in a fundamental way. The IPCC's Second Assessment Report (1996) contains an explicit discussion of this relationship. Yet, amazingly, the IPCC's review of work on the value of this parameter suggests values that are *substantially larger than the values that have been estimated by most investigators*. According to the IPCC (1996),

> Although no consensus has emerged [on the value of the elasticity of the marginal utility of consumption], there is a generally accepted method for approaching the issue. . . .
>
> Individuals in their day-to-day decision making reveal information about their perceptions concerning their own utility functions in at least two different contexts: behaviour towards risk and behaviour towards intertemporal allocation of consumption. In both contexts, there seems to be a consensus that elasticities of marginal utilities lie in the range of 1 to 2, even though the empirical studies require strong assumptions about the specific form of the utility function (symmetric and time separable). Thus, one of the most commonly used utility functions, the logarithmic, implies [an elasticity of the marginal utility of consumption] = 1, meaning that if income rises by 1% the marginal utility of consumption falls by 1%. Attempts by Fellner (1967) and Scott (1989) to estimate this elasticity place it somewhat higher, at 1.5, whereas recent estimates reviewed by Pearce and Ulph (1994) place it in the vicinity of 0.8.   (p. 136)

Given the inverse relationship between the intertemporal elasticity of substitution and the elasticity of the marginal utility of consumption, this implies a range for the intertemporal elasticity of substitution from 0.67 to 1.25. (For Cobb–Douglas utility, $\sigma = 1$.)

Yet macroeconomists (the group of economists who have examined the intertemporal elasticity of substitution most extensively) have estimated a range of values for σ from essentially zero to around 0.3. For example, Hall's (1988) careful econometric analysis of how aggregate consumption responds to changes in real interest rate expectations is unambiguous: "[A]ll the estimates presented in this paper of the intertemporal elasticity of substitution are small. Most of them are also quite precise, supporting the strong conclusion that the elasticity is unlikely to be much above 0.1, and may well be zero" (p. 340). Kocherlakota (1996) shows that it is possible to restate the famous "equity premium puzzle" (Mehra and Prescott 1985) as evidence that the coefficient of relative risk aversion[26] must be greater than 8.5. This is equivalent to saying that the intertemporal elasticity of substitution is less than 0.12.[27] Approaching the problem with an entirely different methodology – "survey responses to hypothetical situations constructed using an economic theorist's concept of the underlying parameters" – Barsky et al. estimate average lower and upper bounds on the elasticity of intertemporal substitution to be 0.007 and 0.36 respectively, with a midpoint of 0.18 (1997, p. 566).[28] As noted above, Auerbach and Kotlikoff (1987) use a value of $\sigma = 0.25$.

It is a mystery how could the IPCC could have ignored all the evidence and controversy surrounding the empirical value of this key parameter. The IPCC's statement itself reflects a high degree of confusion – first stating that there is no consensus, then asserting that the consensus is a value of the elasticity of the marginal utility of consumption between 1 and 2. It does not help that the IPCC notes that "for one of the most commonly used" functional forms, the logarithmic, the crucial elasticity is one; the elasticity of the marginal utility of consumption (and of intertemporal substitution) is one *by assumption* if logarithmic (or Cobb–Douglas) utility is specified. Of the three empirical studies cited by the IPCC, one was published in 1967 and one is an unpublished mimeo. This can hardly be construed as indicative of a consensus given the hundreds of articles that have been published in recent years on individuals' attitudes towards risk, the equity premium puzzle, and related topics, as well as the more recent and statistically rigorous estimates of the intertemporal elasticity of substitution itself.

The intertemporal elasticity of substitution is not the only critical parameter. The models worked out in this chapter take all the substitution parameters to be the same, but this surely is not descriptive of the real world. To go back to the point made in Chapter 2, when the question involves the substitutability of ordinary goods for macroenvironmental goods, it is difficult even to imagine a methodology that could provide reliable empirical estimates of the elasticity of substitution between the different types of goods. The climate catastrophes some scientists fear have not occurred, and are unimaginable to most citizens. No empirical information exists that would enable reliable estimates to be made of how such events might enter individuals' utility functions.

These issues may be at the crux of the disagreements between "ecological economics" and "mainstream economics." If one is committed to the proposition that environmental changes are gradual and linear, and that produced services are easily substitutable for natural services, then one will gravitate towards the "mainstream" position that we have plenty of time to address greenhouse gas (GHG) emissions, that climate damages have to be compared in a marginal way to the benefits of other production forgone, and so forth. On the other hand, if one believes that produced goods and natural services are not easily substituted for each other, that macroenvironmental changes are likely to be large and nonlinear, and that the time pattern of environmental change is very important, then the "ecological economics" perspective feels more like the nature of reality.[29]

This leads to the final, essential, point: what matters most in climate policy is the answer to the question, "Who is at the table?" The flaws in current integrated assessment models stem from their insisting that all analysis be carried out with a present-oriented stance. Whether this is done by discounting future costs and benefits, or by imposing restrictions on utility functions that place lower value on the future, or by specifying that there is a high degree of substitutability between present and future goods (and between environmental and ordinary produced goods), the result is the same. What appears to be a scientific analytical exercise carries within it a heavy burden of implicit ethical choices. The details of the analytical models are less important than their uniformity in imposing a present orientation. This has nothing to do with the notions that there is a declining marginal utility of consumption, or that investments are productive, or any of the other standard assumptions of neoclassical analysis. General equilibrium theory in itself is sufficient to generate an "environmentalist" policy orientation if the models are set up in such a way as to make sure that all agents are present and represented in the market.

Of course, in practical terms the representation of people not yet born can only take place through applications of *principles*. How to do this should be a major item on the research agenda for climate policy analysis, and for the political discussion of climate issues. General equilibrium theory as we currently know it can provide some guidelines, both positive and negative. We can see what equilibria look like when people from different time periods engage in model transactions, and we can discern the consequences of imposing a present-oriented slant on model design. It is clear from the examples discussed in this chapter that economic logic shows no preference for individuals of different time periods; the present-centric bias that comes with ordinary discounting is an artifact of incomplete models or hidden assumptions. It is difficult to translate the outcomes of hypothetical general equilibrium models into working principles for policy design. Even so, development of these principles should be a focus of research and political discussion. Real progress in the climate debate has first to resolve the question of whose interests are being represented, through a transparent and complete specification of how various kinds of rights are to be assigned. Then the power and elegance of general equilibrium analysis can be brought to bear to explicate the nature of the choices we face. Current practice only hides the essential questions behind a technical facade.

4
The Representation of Production

4.1 Introduction

"Tastes" and "technology" are the fundamentals upon which neoclassical general equilibrium models are built. While the representation of tastes is expressed through utility and demand functions, the underlying productive technology of the economy is represented by production functions that relate various quantities of various inputs to the output produced by firms. This approach requires a behavioral theory that describes the activity of firms as they carry out the processes of production. The behavioral theory is profit maximization, which parallels the utility maximization of consumers on the demand side of the economy. While some of the conceptual and mathematical difficulties that plague the representation of preferences are common to the treatment of production in the models, there are several important differences.

Because of the way firms are defined, there are no wealth effects that could create the kind of multiple equilibria that we have seen in exchange models that embody consumer preferences.[1] Consumers maximize utility subject to their budget constraints, which, as in the models of Chapters 2 and 3, depend on the endowments of the different kinds of goods over which the consumers have property rights. In the neoclassical representation of the firm, the firm has no endowments – it simply hires factors of production and maximizes profits, subject to its technology of production. Because it has no endowments of inputs, there can be no wealth effects associated with changes in relative input prices. As a consequence, the factor demand curves are invariably downward-sloping functions of the factor's "own price."

Mathematically, imagine a firm that produces a single output with a market price p. (There is no difficulty in generalizing to multiple outputs.) The conventional assumption (as in Chapters 2 and 3) of perfect competition will be maintained throughout. Then, if the factors of production are $x_1, x_2, \ldots, x_m$ with prices $w_1, w_2, \ldots, w_m$, and the production function transforming the inputs into output is f, the firm maximizes profit

$$\pi = pf(x_1, x_2, \ldots, x_m) - \sum_{i=1}^{m} w_i x_i \tag{4.1}$$

The first-order conditions for a maximum (with $x_1^*, x_2^*, \ldots, x_m^*$ being the equilibrium values) are

$$pf_i\left(x_1^*, x_2^*, \ldots, x_m^*\right) = w_i \quad \forall i \tag{4.2}$$

There are no endowments in equation (4.2), hence the absence of any of the wealth effects. Indeed, it is easy to show that, provided the second-order conditions guaranteeing a maximum hold, the "own price" factor demand curves are downward sloping (Silberberg 1990), that is

$$\frac{\partial x_i^*}{\partial w_i} < 0 \quad \forall i \tag{4.3}$$

Similarly, it is very easy to show that equilibrium profits cannot increase if the price of any factor of production increases. Thus, if π^* is the maximum profit for a particular set of factor prices,

$$\pi^* = pf\left(x_1^*, x_2^*, \ldots, x_m^*\right) - \sum_{i=1}^{m} w_i x_i^* \tag{4.4}$$

then for any particular input j,

$$\frac{\partial \pi^*}{\partial w_j} = p \sum_{i=1}^{m} f_i\left(x_1^*, x_2^*, \ldots, x_m^*\right) \frac{\partial x_i^*}{\partial w_j} - \sum_{i=1}^{m} w_i \frac{\partial x_i^*}{\partial w_j} - x_j^*$$

$$= \sum_i \left[pf_i\left(x_1^*, x_2^*, \ldots, x_m^*\right) - w_i \right] \frac{\partial x_i^*}{\partial w_j} - x_j^*$$

$$= -x_j^* < 0 \tag{4.5}$$

because each of the terms in the bracketed summation in (4.5) is zero from the first-order conditions. An argument such as that embodied in equation (4.5) is the reason for many economists' skepticism about the "Porter hypothesis" – the notion that stricter environmental regulation can, under some circumstances, have a beneficial effect on corporate performance and profitability. Tighter environmental regulation is, in the conventional view, equivalent to an increase in the price of one of the "inputs" to production, where the input is considered to be a pollutant and the price is the cost that is imposed, implicitly or explicitly, by the environmental regulation. If the firm is optimized and the combination of the behavior and technology or production is described in the framework of equations (4.1)–(4.4), then the increase in environmental stringency cannot increase profitability.[2]

Once the "problem of production" is stated in this manner, however, several difficulties are apparent. First, some firms actually do possess endowments of goods. Oil companies are one example. This is not really a valid objection to the framework of equations (4.1)–(4.4) because it could be argued that the companies had to pay for their rights to the oil, or, more fundamentally, that the actual owners of the resource are the shareholders of the company, and that the task of the management of the company is simply to maximize profits given the firm's technology. "Rents" would have to be paid to the owners of the oil no matter who they were, or conversely, the owners of the oil would sell it at its market price regardless of whether "their" firm or some other were the buyer.

A more serious objection arises in the treatment of technological knowledge as something that can be adequately represented by a production function transforming commodity inputs into output. While such a representation might be sensible under unchanging pastoral conditions, it is far less clear that it is appropriate in the modern world characterized by rapid technological change. Furthermore, what of the behavioral model? We know that the managers of firms do not spend their time solving the maximization problem (4.1); they are instead engaged in a multitude of activities of varying levels of difficulty, requiring an entire range of human capabilities from leadership to the application of the insights of mathematical finance. At the most fundamental level, the production function and profit maximization representation of the firm gives no insight into why firms exist at all. How are the boundaries of firms determined? Why do actual business organizations carry out some activities themselves, while contracting out other functions and purchasing yet others in the market?[3]

The mathematics of (4.1)–(4.4) is simple and its implications are straightforward. What, then, of the myriad of evidence that firms display inefficiencies and experience difficulties of all sorts? Why are firms so diverse, to the point that some prosper and others fail? Why are top managers so highly compensated? We shall see that deep deficiencies in the production function/profit maximization model arise from a number of sources. Yet despite these red flags, the practice in the economic models used for climate policy analysis is universally to represent production in some (often quite elaborate) version of the production function/profit maximization model.[4] The consequences of this practice are very strong, and the result is an upward bias in estimates of the costliness of climate protection measures. As in the case of consumer preferences, the subtleties of the behavior of firms as they operate in the real world are well-known both within and outside the economics profession. What is missing is a realistic application of those insights to the construction of climate policy models.

4.2 The modern theory of the firm

Suppose we begin by accepting, in the spirit of the "inside critique" of Chapter 2, that firms and other productive organizations are collections of rational, self-interested individuals. At the foundation of the production function/profit maximization story is the notion that the firm is *efficient*. That is, profits are *maximized* given the technological possibilities embodied in the production function. The consequence of this maximization is a series of first-order conditions equating factor prices to marginal products. These equalities in turn form the starting point for econometric estimation of production and supply functions and of comparative statics calculations of how the firm would respond to changes in market conditions. (The "proof" in equations (4.4)–(4.5) that profits cannot increase if environmental regulations are tightened is an example of such a comparative statics exercise.)

The most basic insight of the modern theory of the firm is that the individual rationality of the members of the firm does not guarantee, and in fact is likely to be a barrier to, the "rationality" of the organization, defined as optimal profit maximization. The interests of the individuals making up an organization are not identical to the interest of the organization as a whole. There are substantial areas of overlap in interests, of course – in the extreme, if a firm goes out of business its members are unemployed[5] – but the task of aligning the actions of the members of the organization with the organization's formal goal is a formidable one.

This is a leading example of the well-known problem of collective action. The wide scope of this problem was set out with great clarity by Olson (1965), but awareness of it has been with us since the earliest times. The precise nature of the divergence between the interests of the

individuals making up a group and the interest of the group as a whole
has been described in various ways, as the principal/agent problem, the
problem of opportunism, and the problem of information asymmetry.
All these descriptions are ways of looking at similar sets of difficulties,
so it is worthwhile to discuss some of the specifics of the argument.
There is an extensive technical literature spanning economics, management science, sociology, and organization theory that covers these
issues in detail, and no attempt will be made here to survey that literature comprehensively.[6] Instead, the main arguments will be reviewed
in somewhat generic form. What is common to all these literatures is a
consensus that the simplistic production function/profit maximization
framework does not and cannot serve as an adequate description of the
activities of firms and other productive organizations in a modern
economy.

4.2.1 Principal–agent problems

All human relationships have at least the potential for a divergence of
the interests of the parties. While cooperation is necessary for any kind
of complex production or market interaction, it should not come as a
surprise that the leaders (or owners) of firms must struggle with how to
induce the members of the firm to advance the firm's objective – which
is, at the most basic level, to increase the owners' wealth. The conflicts
that can come into play in relationships of this type (which are not confined solely to the modern business corporation) are referred to as "principal–agent problems" because the interests of the "principal" (who may
be the shareholders of a firm, the top management, or some subgroup
of the organization) and the "agent" (the employees of the firm, a
management layer below the top, or a subordinate subgroup) do not
coincide.

Some principal–agent problems arise simply because of conflicts of
interest. Owners of a firm may wish that their employees would put
profit maximization above all other goals, but it is impossible to specify
in detail every eventuality in which management discretion is required.
That is, after all, one reason for hiring managers in the first place. In
the natural course of events, situations are bound to arise in which the
managers' interests and those of the owners differ. The consequences
of this "separation of ownership and control" were remarked upon by
Berle and Means in the 1930s, but recognition of the problem goes
back to Adam Smith and before (Berle and Means 1932; Jensen and
Meckling 1976).[7]

Managers are very likely to be privy to information that is unavailable to the owners of a firm. Under such circumstances, the owners have no way of assessing whether the managers' decisions are based on information only the managers have, or are being driven by some distinct interest of the managers. Is the construction of opulent corporate offices necessary to maintain the company's image and to provide a kind of advertising (or signaling), or does it merely flatter the vanity of management? Is entering a new line of business justified on strategic grounds of profit potential, or is it the managers' empire building? Is a corporate jet necessary to economize on scarce managerial resources, or is it a perquisite that could be dispensed with to the benefit of the bottom line? The very disconnection between the advancement of capital on the one hand, and immersion in the day-to-day operations of the firm on the other, that makes modern market capitalism possible is also the source of this kind of information gap. The inability of the principals to know the actual rate of return on various investment projects may induce them to set a "hurdle rate" higher than the appropriate risk-adjusted cost of capital, as a way of assuring that profits are not dissipated into managerial slack (Antle and Eppen 1985). This means that some genuinely profitable projects will not be undertaken, and that the firm's production process will be inefficient as a result.

Principal–agent problems also show up in the design of compensation plans for managers and employees. Stock option plans, investment of employees' retirement savings in company stock, and bonuses based on company or division-wide profit performance are all examples of the attempt to conjoin the interests of the agents of a company with those of their principals. These plans can themselves have adverse repercussions for efficiency, as when managers concentrating on short-run results (because that is what determines their compensation) emphasize temporary gains over long-term opportunities (Statman and Sepe 1984; Pinches 1982), or when the compensation plans are nothing but schemes to conceal managers' collection of excessive rents (Bobchuk et al. 2001).

4.2.2 Problems of control

Even if a divergence of interests between owners and managers (or between different levels of managers, or between management and workers) did not exist, a modern complex organization would face difficulties of control. What are the instructions to be given to lower management and employees in directing them to maximize profits? The modern firm is not like a Robinson Crusoe sole proprietorship or small family workshop; no one working on the assembly line, on the road making sales, or sitting in a customer service cubicle, can possibly know about all aspects of the firm's multitudinous activities. Even the top management cannot know everything; information is filtered as it rises through the bureaucracy. It surely will not do simply to instruct employees to "maximize profits," because no one can know exactly what any person's or group's contribution to the final economic profit of the firm is. Employees have to be given particular tasks and responsibilities; they have to be provided with information systems that enable them to track their progress (and that enable their performance to be evaluated), and specialization within the organization is important in the same way and for the same reasons it is at the industry or economy-wide level.

As a result, the command to "maximize profits" must be mediated by layers of accounting data, direct and indirect performance indicators, internal and external benchmarks, and judgment of intangibles. Everyone knows the pitfalls of having agents work "to the formula" that has been provided as their performance guideline, regardless of the long-term objective. The net profitability of a firm is the result of a complex of decisions made at different levels of the organization, influenced by an external environment (customers, laws and regulations, market conditions) that can only be known imperfectly. Employees have to be motivated to align their actions with the formal objectives of the firm, and the organization must interact with the larger society – its community or communities, the legal system, and "public opinion."

All of these considerations are only a part of what makes management such a demanding task. If maximization of profits along the lines of equation (4.1) were all that was involved, the ideal manager would be a recent college graduate who had done well in beginning calculus courses. There would be no multimillion dollar salaries for top managers, nor would there be a scramble by students for admission to high-priced MBA programs.

4.2.3 Deeper computational limits

At the most fundamental level, the fact that the firm can only make decisions through its own rules and procedures imposes certain limits on its ability to process information. No matter how capable the individuals making up the firm are (and psychological research, experimental evidence, and introspection suggest strongly that individuals themselves are only boundedly rational), the organization does not have a unitary consciousness and does not act in the same way as an entity with a single will. While some advantages can be gained by drawing on the distributed information and intelligence of the individuals who make up the organization, the decision-making procedures of the organization impose one or more layers of complexity on the formation of decisions. Whether these procedures are thought of as a mechanism for aggregating the preferences of the individuals in the firm (as might be appropriate for an owner-operated firm or a relatively small partnership) or whether the rules emerge from legal, cultural, and historical structures, there can be no assurance that the resulting decisions will accomplish any particular singular objective such as profit maximization.[8]

The nonunitary nature of the firm imposes *computational constraints* on the firm's decision-making that preclude complete maximization. Many of the actual problems firms face, such as sequencing and scheduling, database management, storage and retrieval, or network design, can be formalized in such a way that they can be shown to belong to the class of **NP**-complete problems (Garey and Johnson 1979). Mathematically, this means that there is no known algorithm for solving any of these problems that does not require an amount of computation time that grows faster than "polynomially" in the size of the problem.[9] Informally, it means that these problems are intractably difficult and that in practical applications, such as the operation of a productive organization, the managers will have to settle for approximations that fall short of an optimal solution. It is also known that a number of standard problems in economics, including the formation of "rational expectations" and the solution of standard problems in game theory and general equilibrium, run up against computational limits (see Rust 1996 [revised 1997], DeCanio 1999, and the references cited in these papers). Even if human brains are somehow able to access quantum computational effects, communication over "classical" channels, as required in human organizations, means that problems requiring combining information from the various members of the organization are subject to the limits of a classical Turing machine.

The strong implication is that there is no way around the limits that "bounded rationality" places on human economic activity.[10] In particular, it is impossible to achieve the mathematical ideal of full optimization in production processes and the other activities of firms. Simply recasting these problems as calculus maximization exercises that *can* be solved amounts to abstracting from essential features of the underlying reality. The fact that it is usually possible to calculate the optimal solution of equation (4.1) is not sufficient grounds for asserting implicitly that (4.1) is *in fact* an adequate representation of the technological and behavioral reality of the firm.

4.2.4 The evidence

The issues discussed so far in sections 4.2.1–4.2.3 constitute a compelling set of theoretical reasons why we should not expect optimization to be the rule for firms and other productive organizations. In addition to these abstract arguments, there is a very large body of evidence that is consistent with a failure of complete optimization in production. The first and most obvious type of evidence is the personal experience we all have had in our own organizations – each of us knows of a myriad of inefficiencies close to home. It is not necessary to rely only on subjective experience to establish that inefficiency in production is the rule rather than the exception, however. Several currents in the empirical literature of economics, management science, and operations research are devoted to the measurement of inefficiencies in production. In a series of papers beginning in 1966, Harvey Leibenstein introduced the concept of "X-efficiency" (and also its converse, X-inefficiency) and argued that X-efficiency was more important than allocative efficiency in determining the productivity and profitability of firms.[11] Although Leibenstein is most often acknowledged for his theoretical contribution, it should be remembered that his original 1966 paper contained a great deal of empirical information on the existence and magnitude of X-inefficiency, including data from detailed surveys of industrial establishments, the lag time between invention and innovation (or the adoption of the new technology), and the rate of return to management consulting services.

One way of formalizing the notion of technical efficiency (and of measuring the distance a firm or facility is from its production-possibilities frontier) is through the technique of "data envelopment analysis" (DEA). Originally introduced in the 1950s (Koopmans 1951, Farrell 1957), the DEA technique calculates the efficient points on the isoquant or production-possibilities frontier for a group of individual decision-making units (DMUs). The DMUs that lie on the frontier are "technically efficient," while those that lie inside it are inefficient. In effect, DEA is a flexible method for "benchmarking" DMUs to the most efficient comparable units in their industry.

Several comprehensive standard treatments of DEA are available (Cooper et al. 2000, Charnes et al. 1994, Sengupta 1995). The DEA literature itself is extensive; the CD-ROM bibliography (Seiford 2000) accompanying the Cooper et al. text and covering the period 1978 through September 1999 contains over 1500 entries.[12] Articles containing "data envelopment analysis" in their title, keywords, or abstract have been running at a rate of around 100 per year in recent years in the Social Sciences Citation Index (ISI Web of Science 2002). It would be an interesting (although extremely time-consuming) exercise to conduct a complete meta-analysis of this literature to determine if there are any patterns or tendencies in the levels of inefficiency reported; a nonsystematic compilation of 26 such studies revealed an average efficiency level of 86 percent (DeCanio 1997).

DEA is not the only approach to the measurement of relative efficiency. DEA is a nonparametric technique, but parametric methods have also been applied to the problem.[13] In addition, there are other indications of the failure of profit maximization to account for what we observe in the real world. If firms were fully optimized, there would be no market for corporate control, and no efforts by management to insulate themselves from takeovers by defensive measures such as supermajority provisions and "poison pills." Successful takeover attempts tend to raise the value of the stock of both the target and acquiring firms (Jensen 1988, Bradley et al. 1988, Jarrell et al. 1988), and there is evidence that when states pass antitakeover legislation, the value of firms headquartered in those states falls.[14] All these evidences of the failure of firms to optimize are observationally grounded. It can hardly be considered good scientific practice to ignore such extensive and diverse evidence of the failure of optimization in favor of an unsubstantiated preference for the unadorned profit maximization model of firms and production.

4.3 Aggregation problems

In addition to the deficiencies in profit maximization as a *behavioral* hypothesis, there are also *technical* obstacles to the representation of production by conventional neoclassical production functions. If technological production possibilities are to be represented by any means other than an exhaustive listing of *every* individual piece of equipment along with directions for its use, the production function must embody some notion of capital aggregation. Yet no economically meaningful method of constructing capital aggregates can be devised, except under conditions that are so unusual and special as to have no relevance to the real world.

The basic problem is that the price of capital goods depends on the expected future stream of profits that can be earned through their employment. This future net revenue stream has to be discounted by a discount rate that itself is the price of capital, but such a self-referential pricing problem would not in and of itself preclude a solution (equations involving their own solution as a functional argument can often be solved for their "fixed points"). The real problem is the uncertainty that is intrinsic to technological progress. For it is always possible that new inventions will render elements of the existing capital stock obsolete and worthless. (Recall the example of the electromechanical calculators from Chapter 2.) No measurement of "capital" based on past dollars invested can escape this possibility, and hence the value of the capital stock is always contingent upon the unknowable future state of technology.

This conundrum has been elaborated in a number of different ways. Joan Robinson (1953) made clear that the attempt to reconcile the decision-making of the "man of deeds" (the capitalist business owner making investment decisions based on expected future profits) and the "man of words" (the economist or accountant attempting to measure the capital stock as something other than an exhaustive list of different types of equipment) was impossible under any but the most rigid kind of unchanging conditions. A more recent argument with similar implications is given by Fisher and McGowan (1983; see also Fisher et al. 1983), who demonstrated that the true ("economic") rate of return cannot be recovered from accounting data.[15] As Fisher puts it, "[w]hile the economic rate of return is the magnitude that properly relates a stream of profits to the investments that produce it, the accounting rate of return does not. By relating *current* profits to *current* capitalization, the accounting rate of return fatally scrambles up the timing" (1984, pp. 509–10).

The capital aggregation problem is not just one of measurement. In a series of articles published in the *Review of Economic Studies* (1965, 1968a, b) and *Econometrica* (1969a) and summarized in his definitive *Econometrica* paper (1969b), Fisher describes (and solves, to the extent that it can be solved) the problem of defining aggregate capital stocks and constructing aggregate production functions. In brief, the main results are:

(1) Unless efficient allocation of resources is assumed,[16] an aggregate production function will exist if and only if every firm's production function is additively separable in capital and labor;

(2) If an efficient allocation is assumed, and if labor exhibits strictly diminishing returns, then a necessary and sufficient condition for capital aggregation (and existence of an aggregate production function) is that every firm's production function satisfy a partial differential equation of the form

$$\frac{f^v_{KL}}{f^v_K f^v_{LL}} = g(f^v_L) \quad \forall v \tag{4.6}$$

where the function g is the same for all firms. Here the f^v are the production functions of the individual firms (indexed by v), and the subscripts indicate marginal products or cross-products with respect to the firms' capital and labor inputs. This condition is not easy to interpret directly, but it has two consequences:

(2a) If constant returns is not assumed, "there is no reason why perfectly well behaved production functions cannot fail to satisfy any partial differential equation [of this] form," and "if some firm has such a production function, then exact capital aggregation is impossible regardless of the nature of the production functions of other firms"; and

(2b) If constant returns is assumed, then "*average* product per worker (as well as marginal product) and profits per worker will be the same in all firms" (1969b, pp. 559–60, italics in the original).

Of course, it would be stretching reality past the breaking point to presume that average product per worker and profits per worker are the same in all firms, whether in the whole economy or in any subsector or industry within the economy. If this were not bad enough, the technical conditions for the construction of subaggregates (that is, aggregates comprised of some but not all types of capital) are also very restrictive. Nor is it the case that an aggregate production function with nice properties can be found that closely approximates the true disaggregated production function. The only way in which such approximations could be assured would be if we were willing to accept production functions that are very "irregular" in the sense that their first and second derivatives exhibit large "wiggles" up and down. Fisher is not entirely pessimistic about the possible usefulness of aggregate production functions, but he concludes on this note:

> Just because it is possible to use aggregate production functions for grand statements about long run growth and technical change, it is important to be careful about the foundation for such statements. At

present, that foundation seems solid only insofar as relatively small changes are concerned. The analyses which I have here summarized have convinced me that there is at least need for great caution in this area. It may recalled that Solow's seminal article [1957, p. 312] called for "more than the usual 'willing suspension of disbelief' to talk seriously of the aggregate production function." That suspension has clearly led to very fruitful results. I am, however, finding it increasingly difficult to maintain. The conditions for the existence of aggregate production functions, at least when widely diverse industries are included, seem very, very strong.   (1969b, p. 576)

What are the implications of these findings for current practice in integrated assessment modeling? The models are not based on any explicit tests of whether the conditions required for aggregation of production functions hold.[17] This means that even if the production functions of the models bear some kind of statistical relationship to past data, the production functions cannot therefore be taken to be *empirically grounded*. Instead, they represent a fiction that is mathematically convenient, one that enables the derivation of optimality conditions (expressed as the first-order conditions for profit maximization) that might be useful for policy analysis if they actually did obtain. The plausibility of this fiction may vary according to the perceptions of the consumers of the models' results, but that kind of postmodern attitude (that "reality" is in the eye of the beholder) is at variance with the cloak of scientific rigor in which the models are typically wrapped.

4.4 A more realistic characterization of production

We have seen in Chapter 2 that the mathematical behavior of a tâtonnement adjustment process depends on the characteristics of the market excess demand functions. Without knowledge of the actual dynamics of the economic system, comparative statics analyses of the response of the system to policy changes are not reliable. The excess demand functions derived from standard utility-maximizing assumptions can have virtually any shape, so that the use of equilibrium conditions to characterize the economy leads to an underdetermined system. This suggests that the dynamic processes themselves might be a better starting point for modeling than the conditions of equilibrium. A true dynamic model could trace out the development of the system over time, whether or not it tends towards equilibrium (and regardless of the characteristics of any equilibrium that might exist). If the dynamics were known, no information about the behavior of the system would be lost or missing, whereas we have seen that knowledge of the equilibrium conditions alone is not sufficient to determine the system's characteristics. A focus on dynamics rather than equilibrium may also be a better way to approach the representation of production.

There is no doubt that the problem of dynamics is difficult, and has not been satisfactorily solved (at least there is no consensus about what constitutes a good dynamic economic model). One way of addressing production dynamics that has much to recommend it is an *evolutionary* perspective. Evolutionary models have the advantage of being well-suited to complex systems, and the pressure of "natural selection" is highly analogous to the market pressures faced by firms in a competitive environment. Evolutionary thinking about economic dynamics has a long and distinguished history. Hayek (1988) argued that the idea of evolutionary change in the social sciences and humanities predated its acceptance in biology, although the evolution of cultures and institutions has to be based on the inheritance of learned characteristics, unlike the mechanisms of biological evolution. Milton Friedman (1953) appealed to an evolutionary dynamic in making the argument that whatever their apparent behavior, firms behave "as if" they are maximizing profits, because otherwise competitive pressures would drive them out of business. Unfortunately for Friedman's argument, however, there is nothing in evolutionary theory that requires that natural selection achieve *optimality*; the reality of the biological world is that variability characterizes the fitness of both species and individuals in a population, and that environmental change and coevolution can render past adaptations obsolete. Interestingly, a very clear exposition of the power of the evolutionary argument in economics, which avoids the errors and misconceptions of Friedman's essay, was published several years before Friedman by Armen Alchian (1950). It is worth quoting Alchian at length:

> Current economic analysis of economic behavior relies heavily on decisions made by rational units customarily assumed to be seeking perfectly optimal situations. Two criteria are well known – profit maximization and utility maximization. According to these criteria, appropriate types of action are indicated by marginal or neighborhood inequalities which, if satisfied, yield an optimum. . . . [I]t is alleged that individuals use these concepts implicitly, if not explicitly. . . .

There is an alternative method which treats the decisions and criteria dictated by the economic system as more important than those made by the individuals in it. By backing away from the trees – the optimization calculus by individual units – we can better discern the forest of impersonal market forces. This approach directs attention to the interrelationships of the environment and the prevailing types of economic behavior which appear through a process of economic natural selection. Yet it does not imply that individual foresight and action do not affect the nature of the existing state of affairs.

In an economic system the realization of profits is the criterion according to which successful and surviving firms are selected. This decision criterion is applied primarily by an impersonal market system . . . and may be completely independent of the decision processes of individual units, of the variety of inconsistent motives and abilities, and even of the individual's awareness of the criterion. The reason is simple. Realized positive profits, not *maximum* profits, are the mark of success and viability. It does not matter through what process of reasoning or motivation such success was achieved. The fact of its accomplishment is sufficient. This is the criterion by which the economic system selects survivors: those who realize positive profits are the survivors; those who suffer losses disappear.

The pertinent requirement – positive profits through relative efficiency – is weaker than "maximized profits," with which, unfortunately, it has been confused. . . .

(pp. 211–13, emphasis in the original, footnotes omitted)

It is clear that Alchian appreciated the difference between selection pressures that favor the more profitable firms and the much stronger assertion that these selection pressures result in the survival of only producers whose techniques are *optimal*. As alluded to above, the empirical literature overwhelmingly supports the conclusion that populations of firms exhibit a *distribution* of profitability or efficiency, rather than all exhibiting complete maximization.

The most highly developed synthesis of evolutionary economics is the theory of Nelson and Winter (1982). Their insights have been developed and extended since *An Evolutionary Theory of Economic Change* first appeared,[18] but the comprehensiveness of their critique of neoclassical orthodoxy – and the depth of their insight into how to reformulate economic theory – has never been surpassed. In discussing what is at stake in the contest between evolutionary and standard neoclassical models of the firm, they observe

The general issue here is this. A historical process of evolutionary change cannot be expected to "test" all possible behavioral implications of a given set of routines, much less test them all repeatedly. It is only against the environmental conditions that persist for extended periods (and in this loose sense are "equilibrium" conditions) that routines are thoroughly tested. There is no reason to expect, therefore, that the surviving patterns of behavior of a historical selection process are well adapted for novel conditions not repeatedly encountered in that process. In fact, there is good reason to expect the opposite, since selection forces may be expected to be "sensible" and to trade off maladaptation under unusual or unencountered conditions to achieve good adaptations to conditions frequently encountered. In a context of progressive change, therefore, one should not expect to observe ideal adaptation to current conditions by the products of evolutionary processes.

(1982, p. 154)

Economics and the social sciences are still a long way from having agreed-upon formal models of evolution and dynamics. It is possible, however, to indicate how the discovery process might proceed by giving examples of how different evolutionary models correspond to alternative dynamic stories, and how simulations of those models could be compared to what might be observed. It may be ambitious to suggest that a "dynamics first" reconstruction of economic theory is capable of improving on the results of equilibrium theory, but the insight that would be provided by such a theory, together with the known inadequacy of equilibrium theory, calls for making the effort.

4.4.1 Modeling alternative evolutionary dynamics of production

A suitable model has to have two elements: (1) specification of the firm and the nature of the decision problem facing the firm, and (2) specification of the evolutionary mechanism through which selection pressure shapes the outcome of the market process. The illustrative model that will be described below draws on earlier work that radically simplifies the behavior of the individuals making up the firm (and thereby eliminates principal–agent problems) and focuses instead on the firm's problem of selecting its organizational structure (DeCanio et al. 2000, 2001).[19] The model of the firm has been described in these papers and will only be sketched here, in order to focus on the comparison of different evolutionary dynamics.

Formally, the firm is represented as a digraph or directional graph $G(V, E)$, where $V = \{v_1, v_2, \ldots, v_n\}$ is a set of n "vertices" or agents making up the firm, and E is the set of ordered pairs $E = \{(v_i, v_j), \forall i, j \text{ with } i \neq j\}$, representing the "edges" or directed connections between the agents.[20] The organization operates over a discrete series of time steps, and the job of each member is simple: to decide in each time step whether to adopt or not adopt a particular profitable innovation. The decision rule of the agents is as simple as possible: during each time step, the agent adopts the innovation if it "sees" another agent who has adopted it, otherwise the agent does not adopt. Agent i "sees" agent j if the directed edge (v_i, v_j) exists. The innovation initially appears randomly to one of the members of the organization. Once adopted, the innovation bestows a fixed benefit A on the agent at the end of the time step in which it is adopted. Thus, if the discount rate is r and the innovation first appears to agent s, the benefit to agent s is $A/(1 + r)$. The innovation will diffuse through the organization according to its network structure. If τ_i is the time step in which agent i adopts (so that $\tau_s = 1$), the total benefit to the firm is

$$B(s) = \sum_{i=1}^{n} \frac{A}{(1+r)^{\tau_i}} \tag{4.7}$$

If connections between the members of the firm were costless, then obviously the highest benefit would be obtained by having everyone connected to everyone else in both directions. This is not possible in the real world, of course, and in the model the cost of connections is represented by having each agent incur a cost in each time step depending on the number of other agents it sees. If the out-degree of agent i is ε_i, then the cost attributable to agent i is

$$C_i = \sum_{j=1}^{\infty} \frac{c^{\varepsilon_i t}}{(1+r)^j} = \frac{c^{\varepsilon_i}}{r} \tag{4.8}$$

This cost function is nonlinear in out-degree, reflecting "information overload" if an agent tries to collect and process data from too many sources at once. Profit or net benefit to the firm when agent s is the initial adopter is

$$\pi(s) = \sum_{i=1}^{n} \frac{A}{(1+r)^{\tau_i}} - \sum_{i=1}^{n} \frac{c^{\varepsilon_i}}{r} \tag{4.9}$$

Because the site of the initial appearance of the innovation is random, the total value of the firm G is obtained by averaging over all the potential initial nodes, so that

$$\pi_G = \frac{1}{n}\sum_{s=1}^{n}\pi(s) \tag{4.10}$$

An illustration of the diffusion of the profitable innovation through a firm of this type is given in Figure 4.1 for a particular firm of size 6. The agents are labeled 1 through 6 in the upper left graph. The innovation is adopted during time step 1 by agent 1, and the adoption is indicated by changing the vertex representing that agent from a circle to a black dot. In time step 2, the innovation is adopted by agents 5 and 6 who "see" agent 1. In each successive time step, the innovation is adopted by those agents who see one that has already adopted. For this case of the initial adopter, the innovation spreads to all agents in five time steps. When the innovation is first adopted by agent 1, the contribution to the firm's profitability is

$$\pi(1) = \frac{A}{1+r} + \frac{2A}{(1+r)^2} + \frac{A}{(1+r)^3} + \frac{A}{(1+r)^4} + \frac{A}{(1+r)^5} - \frac{5c+c^3}{r} \tag{4.11}$$

from (4.9). Note that if the innovation had first been adopted by agent 6, it would not spread at all, because none of the other members of the firm "see" that agent. The cost component of $\pi(6)$ would be the same as in (4.11), but the benefit would only be $A/(1 + r)$. The total value of the firm would be obtained by averaging the expressions similar to

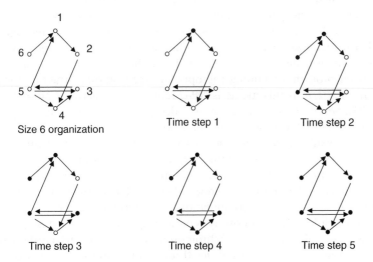

Figure 4.1 Diffusion of innovation through an organization

Table 4.1   Number of digraphs with n vertices

n	Labeled digraphs	Nonisomorphic unlabeled digraphs
1	1	1
2	4	3
3	64	16
4	4096	218
5	1.049×10^6	9608
6	1.074×10^9	1.541×10^6
7	4.398×10^{12}	8.820×10^8
8	7.206×10^{16}	1.793×10^{12}
12	5.445×10^{39}	1.137×10^{31}
24	1.474×10^{166}	2.376×10^{142}

Source: Amir-Atefi (2001). The method for enumeration of the number of nonisomorphic digraphs is given in Harary (1969).

(4.11) that would be obtained from the innovation's initial appearance at each of the nodes.

The managerial decision problem of the firm is to find the network structure that maximizes π_G. Although it is easy to find the optimal structure for particular parameter values (A, r, c) for very small organizations of size 2 or 3, the problem cannot be solved by brute force search for larger values of n, because the number of potential structures increases very rapidly with n. The number of labeled digraphs (in which each agent is distinctly named) and nonisomorphic unlabeled digraphs[21] for the first few values of n are given in Table 4.1.

Given this size of the search space, it is plausible that the managers of real organizations would adopt heuristic methods to approach or approximate the optimal solution. It has been conjectured that the general problem of finding the optimal organizational structure for a firm of size n in this model is **NP**-complete or harder (DeCanio et al. 2001). Some types of organizational design problems are known to be harder than **NP**-complete (Papadimitriou 1996).

The organizational model (4.7)–(4.10) is well-suited for modeling evolutionary dynamics because there is a compact encoding of the organizational structure that is ideal for application of genetic algorithms (GAs) (see Holland 1975, Goldberg 1989, Mitchell 1996 for standard treatments of the GA computational technique). In essence, a GA mimics biological evolution by representing the entities whose population is evolving by their "chromosomes," allowing the creation of "offspring" of the parent entities through mutation and exchange of

Parent 1	*1 0 0 0 0 0 1 0 0 0 0* 0 1 0 0 0 1 0 0 1 0 1 1 0 1 0 0 0 0
Parent 2	1 1 1 0 0 1 0 1 0 0 0 1 *1 0 0 1 0 0 1 0 0 1 0 0 1 1 1 1 0 1*

Offspring	*1 0 0 0 0 0 1 0 0 0 0 1 0 0 1 0 0 1 0 0 1 0 0 1 1 1 1 0 1*

Figure 4.2 Chromosomes of two parents and offspring, size 6 organization

genetic information (analogous to what happens in biological sexual reproduction), and subjecting the entire population to selection pressure over time. In the organizational model, the chromosome is formed by concatenating the rows of the *adjacency matrix* of the organizational structure,

$$\mathbf{A}_G = [a_{ij}] \tag{4.12}$$

where the element a_{ij} of the adjacency matrix equals 1 if the edge (v_i, v_j) is present and equals 0 if (v_i, v_j) is absent. For an organization of size n, this produces a chromosome of length n^2. (In practice, the chromosome length can be reduced to $n^2 - n$ because the diagonal elements of $\mathbf{A}_G$ are all zero and can therefore be omitted.) Fitness of the organization is determined by its profitability according to equations (4.9)–(4.10). Two members of the evolving population can produce an offspring by "crossover" exchange of genetic information. This is illustrated in Figure 4.2 for two parent organizations of size 6.

Parent 1 is the organization shown in Figure 4.1. (Keep in mind that the all-zero diagonal of the adjacency matrix has been eliminated in this representation.) Parent 2 is another size-6 organization. If these two parents are selected to reproduce, and the single "crossover" point is such that the first 12 genes of parent 1 are combined with the last 18 genes of parent 2 (these are shown in boldface italic type in Figure 4.2), the resulting offspring organization is as shown. The GA takes a population of organizations, differentially selects members of the population for reproduction according to their relative fitness, creates offspring by the process illustrated in Figure 4.2, and repeats the steps in each successive generation. In addition, mutations can occur at random as genes are switched from 1 to 0 or from 0 to 1 according to a probabilistic rule.

This algorithm simulates natural selection. It allows "experimentation" in the formation of organizational structures because different pieces of parent organizations can be broken off and rejoined, with the offspring's survival dependent on the resulting fitness of the newly formed structure. Different dynamics can be simulated by changing the rules of the GA. The rules that were varied in the runs reported below had to do with selection for reproduction and whether or not "elite" (high fitness) organizations were preserved from generation to generation. This setup is flexible enough to simulate the evolution of "mixed" populations containing organizations of different size.

4.4.2 Some indicative results

Genetic algorithms are commonly employed to attack difficult optimization problems such as the choice of organizational structure. In the present context we are interested not so much in finding the best structure, but in the characteristics of populations that are evolving under the pressures of natural selection. In particular, the goal is to see how variants of the basic GA technique give rise to populations with different characteristics. By examining these simulated populations, it may be possible to imagine how proper specification of the GA rules could mimic the kinds of populations of firms (varying as to the distribution of productivity or profitability, the size distribution, etc.) that are observed in the real world.

We begin with a population consisting of S "species" of firms. These species are defined solely in terms of the size of the firms. (Size is defined as the number of agents making up the firms.) There will be M members of each species in the initial population, so that the total population size is $M \times S$. The parameters of reward, cost, and discount rate (A, c, r) are fixed. Three selection methods will be considered: proportionate selection, tournament selection, and population elite selection.

Population elite selection is the simplest of the three, and corresponds most closely to the implicit population dynamics of Hannan and Freeman's classic *Organizational Ecology* (1989). Under this selection method, every member of the population has an equal probability of being selected for reproduction. For each species subpopulation, form a number of offspring equal to the size of that subpopulation, choosing the parents at random without replacement and using two-point crossover (which produces two offspring from every pair of parents).[22] Combine all these subpopulations (including both parents and offspring), and select the top 50 percent (as measured by their fitness). Repeat for successive generations. This method amounts to a more or less random formation of new organizations that imitate components of existing organizations; however, there is no conscious selection of the more successful organizations in a population to imitate. Organizations simply are born and die, with survival determined by relative fitness. Selection pressure operates on the population as a whole.

Tournament selection is more sensitive to the fitness of the individual organizations in determining whether they reproduce. Beginning with the existing population, two members are selected randomly (with replacement). Of the two selected, the one with the higher fitness is chosen as the first parent. After that selection is made, two members of the first parent's species are selected at random, again with replacement. The second parent is the one of this pair with the higher fitness score. The two parents chosen in this way then reproduce via two-point crossover with a 50 percent probability. If the crossover does not occur, both parents are copied into the next generation. This method differs from population elite selection in that the candidates for parenthood are chosen in a way that favors the more fit members of the population. This corresponds to the notion that the newly formed firms tend to imitate elements of the more successful existing firms. The reason such a selection process cannot be assumed to take place, of course, is that it cannot be assumed that entrants necessarily know the profitability of existing firms. On the other hand, if the new organizations are formed as a result of spinoffs or joint ventures by existing firms, it is plausible that the more successful firms would have a higher probability of engaging in the formation of new firms. Thus, there is no a priori presumption of which selection dynamic is more plausibly a reflection of competitive market pressures.

The third selection method, proportionate selection, is similar to tournament selection except that the parents are selected for reproduction with probabilities proportional to their relative fitness in the whole population.[23] This method is analogous to the "replicator dynamics" that appears in the literature of evolutionary game theory (Weibull 1995). It represents a different way of favoring the reproduction of the more fit members of the population. Indeed, each of these selection methods can be thought of as selecting population members for reproduction according to some probability distribution. Population elite selection is equivalent to using a uniform distribution, tournament selection is equivalent to a distribution based on rank ordering of the fitness scores, and proportionate selection is equivalent to a distribution taking account of the relative magnitudes of the fitness scores.

With proportionate or tournament selection, the evolution can take place with or without the preservation of population elites. If elites are preserved, some selected number of the most fit members of the population are simply copied over from one generation to the next. This prevents degradation of highly fit organizations; if the GA has found highly efficient structures, the preservation of elites prevents those structures from being broken up through crossover. On the other hand, even successful organizations are subject to "reinvention," internal reorganization, or change in response to fashions and trends. Thus, runs with and without elite preservation were carried out.

A basic question is whether these evolutionary dynamics generate similar population distributions, for different runs of the same evolutionary mechanism and across the different mechanisms. Consider first whether the same mechanism leads to the same distribution of firms for different runs. The particular distribution of fitness scores in any evolutionary run will depend on (1) the initial randomly generated population that provides the basic genetic material for evolution of the population, and (2) the (partially) random events that occur during evolution as crossover and mutation take place. Selection pressure will tend to preserve the more fit structures, but the question is whether the distribution of fitness scores tends to be similar after each evolutionary run. This has the same flavor as the question asked in connection with biological evolution, "What would be conserved if the tape were played twice?" (Gould 1989, Fontana and Buss 1994).

Figure 4.3 shows typical histograms[24] of the fitness scores of two populations evolved under these conditions: firm size 15, population size 500; 10 elite firms saved each generation; crossover probability 0.5; mutation probability (1/210); $(A, c, r) = (500, 3.5, 0.1)$; maximum time steps for diffusion of the profitable innovation = 10. The evolution was allowed to continue for 500 generations, and proportionate selection was the GA type. The first thing to observe from Figure 4.3 is that there is a considerable spread in fitness scores. Even after 500 generations, the bulk of the population falls short of the maximum fitness score achieved by the most profitable organizations. Most of the firms in the evolved populations are clearly not optimized even after 500 generations. It is also evident that the distributions are skewed to the left. This pattern is reasonable because, while selection pressure will promote the survival of the more fit members of the evolving population, the processes of crossover and mutation will at the same time give rise to quite a few failed experiments in organizational structure in every generation. This is akin to the stylized fact that in the real world most new firms fail rather quickly; the rule of thumb is that four out of five new small businesses fail in the first five years of their life.

It is also the case that even for a common set of parameters and a particular GA, the evolved populations do not all exhibit the same distribution of fitness scores. There are a number of statistical tests for whether a group of samples all come from the same distribution. The

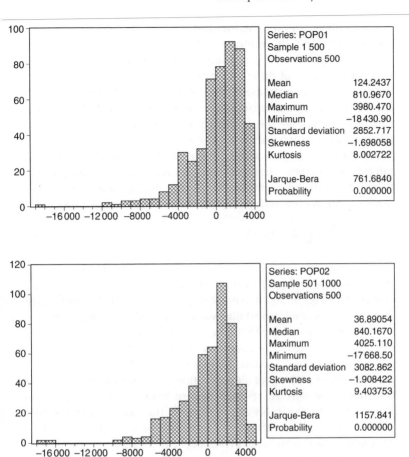

Figure 4.3 Typical histograms of fitness scores from evolved populations

most common is the one-way analysis of variance (ANOVA). However, ANOVA requires that the underlying distribution(s) be normal, and we know that this is not likely to be the case for the fitness scores of either the evolved populations or the original random populations. The fitness function is not the sum of a large number of small independent disturbances; it is highly nonlinear in the out-degree of the vertices (because of the functional form of the cost). Furthermore, the effects of the connections between vertices are not independent influences on total fitness; removal of a single edge can drastically change the fitness of the whole organization if, for example, that edge is necessary to make the organization fully connected.[25]

Even though ANOVA is not suitable for testing whether the population fitness scores of different evolutionary runs come from the same distribution, a nonparametric test such as the Kruskal–Wallis test is appropriate. In this nonparametric version of the analysis of variance, each evolved population is treated as a sample. One hundred evolutionary runs of each variety of GA were carried out, and each population had 500 organizations. These 50,000 observations were then ranked, and the rank sums computed for each sample (run). If there are K samples with rank sums $R_1, R_2, \ldots, R_K$, and n_i observations in each sample (with n observations in all), the Kruskal–Wallis statistic is

$$W = \frac{12}{n(n+1)} \sum_{i=1}^{K} \frac{R_i^2}{n_i} - 3(n+1) \tag{4.13}$$

The distribution of this statistic is well-approximated by the χ^2 distribution with $(K-1)$ degrees of freedom (Newbold 1995; see also the discussion in Siegel 1956). Table 4.2 displays the Kruskal–Wallis statistics and their associated probability values under the null hypothesis that all 100 populations have the same distribution of fitness scores for each GA. Both the original populations and the evolved populations after 500 generations were tested. It is clear from Table 4.2 that in *no* case can the null hypothesis of all populations belonging to the same fitness distribution be rejected at the 5 percent level for the original populations, while in the evolved populations the null hypothesis can be rejected in *every* case with only a miniscule probability of Type I error. This result holds whichever evolutionary mechanism is in operation.[26]

These statistical results are strong, but the economic implications are of equal significance. Even though the initial populations are statistically indistinguishable (in terms of the distributions of the fitness scores of the firms), the evolved populations show fitness distributions that are quite definitely distinct. This holds even for uniform evolutionary and environmental conditions – the same GA, firm size, and cost, benefit, and discount rate parameters. The evolutionary process is "open-ended" in that the fitness distribution of firms is not determined by the underlying parameters and the dynamic selection mechanism. This means that random events or historical accidents can measurably affect the outcome of the process. There is nothing new about such a characterization of evolution – after all, biological evolution has selected for locomotion by creatures with 1, 2, 4, 5, 6, 8, and "many" legs. Just as there is no single evolutionary outcome for the "equilibrium" number

Table 4.2 Tests of null hypothesis that all 100 populations are from the same fitness distribution

GA	Firm size	Initial generation		500th Generation	
		Kruskal–Wallis	Probability value	Kruskal–Wallis	Probability value
ps, no elites	15	91.808	0.683	4482.200	~0
ps, no elites	16	97.225	0.532	5036.137	~0
ps, no elites	17	106.407	0.287	9253.005	~0
ps, elites	15	121.749	0.060	1425.642	~0
ps, elites	16	93.625	0.633	1343.897	~0
ps, elites	17	107.779	0.257	1187.104	~0
ts, no elites	15	87.883	0.780	2665.949	~0
ts, no elites	16	113.901	0.145	2156.114	~0
ts, no elites	17	77.568	0.945	3016.906	~0
ts, elites	15	75.729	0.960	787.931	~0
ts, elites	16	72.323	0.980	880.114	~0
ts, elites	17	74.281	0.970	662.428	~0
pe	15	108.350	0.245	49946.71	~0
pe	16	81.337	0.902	49968.05	~0
pe	17	90.272	0.723	49683.87	~0

Source: See text. ps = proportionate selection; ts = tournament selection; pe = population elite selection. Kruskal–Wallis statistics distributed as χ^2 with 99 degrees of freedom.

of appendages of life forms, there can be no presumption that any particular distribution of efficiencies of *firms* (including the distribution that happens to exist at the present time) represents an "equilibrium." An obvious consequence is that policy interventions might, by changing the initial conditions or injecting new information into the market, influence the evolutionary path of the population of firms.

A second kind of question has to do with whether, in cases in which the initial population contains firms of different species, the evolutionary process always tends to lead to the dominance of one type of firm. The dynamics of interspecies competition can be much more complicated than it might seem. Figure 4.4 shows the number of each of the three species (firms of size 15, 16, and 17) remaining in the population (total population size 1500) at the end of each generation for a particular set of parameters and one evolutionary dynamic. In Figure 4.4, the dashed line represents the number of size 15 firms in the

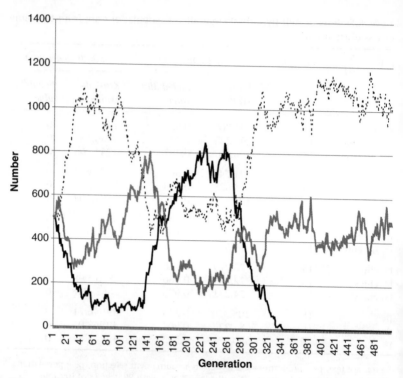

Figure 4.4 15 vs 16 vs 17 proportionate selection with no elites and type 1 scaling

population, the dotted line the number of size 16 firms, and the solid line the number of size 17 firms. It can be seen that the number of the size-17 species drops off precipitously in the first 50 generations, but recovers between the 130th and 230th generations (approximately). There is a stretch of time during this interval when there are more size-17 firms than either of the other two types. However, at around the 250th generation, the number of size-17 firms begins to drop again, until by about the 340th generation these firms have disappeared from the market (the size-17 species has become extinct). After this point, there seem to be about twice as many size 15 firms as size 16 ones. Is this an "equilibrium"? However one might be inclined to answer that question, what is clear is that looking at only a single moment in time could give an entirely misleading impression of the "equilibrium" size distribution of firms. If this industry were examined at the time when the size-17 firms were the most common species, there would be no indication that these firms would eventually be wiped out in competition with the others.

Figure 4.4 gives the result of only one head-to-head competition. To see if there are any patterns in repeated runs, Table 4.3 summarizes the results of 100 such competitions for different versions of the GA. Table 4.3 shows that different outcomes can come about for different evolutionary runs, and that the relative frequency of the different outcomes also depends on the selection mechanism at work. In these runs, the initial population is always equally divided among the three species of firms, and the exogenous parameters (benefit, cost, and discount rate) are identical. The fact that not all the runs with the same selection mechanism come out the same means that pure chance and/or the initial conditions play a role in the outcome. This is the multispecies analogue to the result shown in Table 4.2 that the evolved populations differ under identical dynamics in single-species evolutions. The initial conditions may matter because each run begins with a different population of randomly generated organizations. "Path dependence"[27] may also be playing a role because each generation takes as its starting point the population that had evolved through the previous generation, but the evolutionary outcome is not purely path-determined because of random crossover and mutation that occur along the way.

When proportionate selection with no elites is the evolutionary mechanism, the size-17 firms become extinct about 35 percent of the time. They never dominate the population completely. Only 4 of the 100 runs result in only one species remaining. Most of the runs wind up with a mix of species, predominantly size-15 and size-16 firms. On the other hand, when elites are preserved the proportionate selection runs always lead to a population containing only one species, although which species dominates is a matter of chance (15 percent of the time it is the size-15 firms, 41 percent of the time size-16, and 44 percent of the time size-17). While no elites tends to favor the smaller firms, the preservation of elites favors the larger firms.

If tournament selection is employed, the size-15 firms are almost always eliminated. This form of selection pressure produces populations consisting exclusively of one type of firm, most frequently the size-17 species. Nevertheless, between about a quarter and a third of the runs result in the extinction of the size-17 firms. Once again, chance plays a role in the final composition of the market. Yet for the fixed set of (A, c, r) parameters, there is one size and organizational structure that

Table 4.3 Results of head-to-head competition, 3 species, 100 runs of 500 generations each, $A = 500$, $c = 3.5$, $r = 0.1$

Frequency range	Proportionate selection					
	No elites			Elites		
	Size 15	Size 16	Size 17	Size 15	Size 16	Size 17
0	2	12	35	85	59	56
1–250	0	22	35	0	0	0
251–500	7	37	22	0	0	0
501–750	18	23	6	0	0	0
751–1000	33	2	1	0	0	0
1001–1250	12	1	1	0	0	0
1250–1499	13	0	0	0	0	0
1500	3	1	0	15	41	44

	Tournament selection					
	No elites			Elites		
	Size 15	Size 16	Size 17	Size 15	Size 16	Size 17
0	92	69	39	98	76	26
1–250	0	0	0	0	0	0
251–500	0	0	0	0	0	0
501–750	0	0	0	0	0	0
751–1000	0	0	0	0	0	0
1001–1250	0	0	0	0	0	0
1250–1499	0	0	0	0	0	0
1500	8	31	61	2	24	74

	Population elite selection					
	Crossover probability = 0.5			Crossover probability = 1.0		
	Size 15	Size 16	Size 17	Size 15	Size 16	Size 17
0	100	100	0	100	88	12
1–250	0	0	0	0	0	0
251–500	0	0	0	0	0	0
501–750	0	0	0	0	0	0
751–1000	0	0	0	0	0	0
1001–1250	0	0	0	0	0	0
1250–1499	0	0	0	0	0	0
1500	0	0	100	0	12	88

Source: See text.

has the highest profitability; it is just that the competitive process is not likely to find it.

Under population elite selection, all the runs result in complete takeover of the population by only one species. If the crossover probability is 0.5, the size-17 species always wins the competition. (Recall that the crossover probability is the probability of exchange of genetic information when two parents are selected for reproduction.) When the parents always "mate" (crossover probability equal to one), the size-17 firms usually drive out the other species, but occasionally (12 percent of the time) size-16 species prevail over the other types.

Thus, for identical environmental parameters, the evolutionary selection mechanism can lead to virtually any outcome in the interspecies competition. Sometimes one species predominates, and sometimes a mix of species remains in the population. In the cases in which one species predominates, it can be a different species depending on the run, even with the same selection mechanism. These simulation experiments demonstrate the open-endedness of the evolutionary process.

4.5 Conclusions and implications for policy

These examples illustrate the principal underlying feature of these evolutionary models: the state of an evolving population at any point in time is not uniquely determined. Different dynamic selection mechanisms can give rise to populations with different properties, both as to the size distribution of firms and the distribution of the fitness (profitability or efficiency) of firms. In addition, any given evolutionary algorithm can lead to populations of firms that differ in size composition and fitness scores. While some systematic tendencies may be observed, the variety of outcomes is sufficiently large as to preclude sharp characterizations of the "end-state" populations. Of course, evolution is not a teleological process, so there is no "end-state" in reality. The term is used here to indicate the evolved population that happens to exist at the present time. Nevertheless, confusion between the existing population and some sort of "optimum" is deeply embedded in the "profit maximization subject to the production function" modeling style.

These findings are consistent with other evolutionary models built along similar lines. For example, Amir-Atefi (2001) shows that because the search space in seeking the optimal organizational form is so large, firms that adopt a "rule of thumb" to explore only over team-based structures can frequently outcompete firms whose strategy is to search over the entire space, even though the team-based forms are perforce suboptimal. Mitchell (2001) shows how an evolutionary process by which agents choose their degree of connectedness may or may not lead to optimal network configurations, depending on the initial conditions and the property-rights features of the costs and benefits of technology adoption. In a model that specifies firms carrying out two distinct tasks (one corresponding to the adoption of a profitable technology, as in equations (4.7)–(4.10), and an abstract version of assembly), DeCanio et al. (2001) show that an evolutionary process leads to a range of efficiency levels, and that organizations having very different structures can show comparably high efficiency levels. All of these models represent components of a research program of "computational organizational demography" based on evolutionary dynamics.[28]

This perspective provides a clear scientific rationale for why firms exhibit a range of efficiency characteristics. The differences originate in the nature of the dynamic market selection process itself. In the neoclassical formulation, it is difficult to see why many or most firms appear to be operating inside their production-possibilities frontiers. The attempt to preserve the optimization paradigm by invoking unobserved "transactions costs" that account for the deviations from optimality is unconvincing. It is also unscientific, because there is no evidence that could falsify the hypothesis – by construction the transactions costs are unobservable. The evolutionary model proposed here, on the other hand, is computationally grounded and entirely transparent. It does have the implication, however, that knowing the characteristics of the population of firms (in any particular industry or across industries) today is not in general sufficient to deduce the evolutionary mechanism that gave rise to it – the same mechanism can lead to different populations, and the same population may be the outcome of alternative evolutionary mechanisms.

One clear implication is that there should be no presumption that the existing technology choices, resource allocations, size distributions of firms in industries, or internal decision-making practices of firms are optimized. Of course, this is just the same conclusion as that reached by the modern theory of the firm, and supported by the evidence, as discussed in section 4.2. A number of policy conclusions follow:

• Initial conditions, particular historical contingencies, and strategic policy interventions may affect the technologies and organizational forms used in production without having any adverse effect on overall economic efficiency;

- The possibility that firms (and the economy as a whole) may be inside their production-possibilities frontiers means that there is not necessarily a trade-off between environmental protection (manifested by, for example, a reduction in greenhouse gas emissions) and improvements in productivity or profitability as conventionally measured;
- The range of effective policy instruments is much broader than what has traditionally been considered in the climate debate. In addition to taxes (or tradable permits) and command-and-control regulations, environmental objectives may reasonably be pursued by voluntary programs, campaigns to increase the salience of environmental values, government demonstration projects, and facilitation of inter-firm and interpersonal networking.[29]

Just as in the case of the representation of consumer demand, the representation of production in the prevailing integrated assessment models abstracts from essential features of the phenomena being modeled. The models that have been used to predict the effects of climate protection policies set up their production sectors in such a way as to preclude much of the variation and potential for change that is in fact an essential part of industrial reality in a market economy. By ignoring such vital facts as the variance in efficiency of firms, the integrated assessment models violate standards of scientific practice. In contrast, the evolutionary perspective shows how, within the requirements for internal consistency and computability, dynamic models can exhibit the kind of open-endedness that actually characterizes the real world.

5
The Forecasting Performance of Energy-Economic Models

5.1 Introduction

The predictive performance of economic-energy models has a direct bearing on their usefulness in the climate policy debate. One of the salient aspects of that debate is that some opponents of action to reduce GHG emissions are willing to make confident assertions about what the "cost to the economy" of various policy proposals (such as compliance with the Kyoto Protocol) would be. Yet at the same time these skeptics doubt the validity of the science that underlies forecasts of global warming, and claim that the observed century-long trend of increasing temperatures is indistinguishable from the "noise" of natural variation. They are willing to base strong policy recommendations on the forecasts of economic/energy models that purport to predict output, employment, prices, and emissions decades into the future, while denying the predictive power of physical science models of climate dynamics.

Such a perspective might be understandable if the foundations of the economic models were solidly established while those of the atmospheric and geophysical models were speculative. The truth, however, is just the opposite: the physical and chemical foundations of atmospheric general circulation models (GCMs) are firmly grounded in experimentally verified theory – the kind of theory that predicts that airplanes can fly, that green plants will convert CO_2 to cellulose and oxygen through photosynthesis, and that heat flows from regions of higher temperature to regions of lower temperature in closed systems. On the other hand, as demonstrated in Chapters 2–4, the economic models that are the source of pessimistic predictions about the cost of emissions reductions are dependent on unverified (or, in some cases, false) assumptions about how firms and individuals behave, unexamined ethical judgments about the relationship between present and future generations, and implicit welfare criteria that inordinately favor the status quo. In other words, those who would put off dealing with climate change are skeptical of science that is fundamentally sound, but uncritically accept economics that is theoretically, empirically, and ethically questionable.

Despite the weaknesses of integrated assessment models that have been detailed in the previous chapters, it is still possible that those models might be useful in climate policy analysis if they do a good job of forecasting either the future course of the economy (including energy supply, demand, and prices) and/or they offer reliable guidance about the quantitative effects of various policy alternatives. Even if the theoretical foundations of the models are suspect, they might still serve as forecasting "black boxes" or as a means of calculating policy scenarios. There is, after all, a strand of economic theory that claims that it does not matter whether firms really do maximize profits or not, or whether consumers maximize well-defined utility functions subject to a budget constraint; it is enough that the firms and consumers act "as if" they do so. This approach to predictive analysis was articulated perhaps most clearly by Milton Friedman in his *Essays in Positive Economics* (1953).

Friedman's approach has been criticized elsewhere (Simon 1963, Laitner et al. 2000; see also the discussion of evolutionary models in Chapter 4). Suffice it to say that the "as if" hypothesis rests on strong presumptions about the underlying dynamics of the economic system, presumptions that have neither been established nor tested. Friedman appeals to an evolutionary argument (if the firms did not act "as if" they maximized profits, they would be driven out of business) that is not a necessary feature of real evolutionary processes. Evolution may exert pressure in the direction of greater efficiency, but it is a nonteleological mechanism that typically allows room for continuous improvement. But even if the "forecasting black box" conception of energy/economic models were acceptable in theory, we would still want to know if the models offer accurate forecasts in practice. In order to place any confidence in the models' predictions of policy effects, we would want to know what their record has been in the past.

5.2 The long-term predictive power of economic models is limited

Anthropogenic climate change is a process that unfolds on time scales measured in decades. Economists since Malthus have understood that it is important to *think* about the long run. But thinking systematically about (and gaining insights into) the long run and the factors that shape it – technological progress, stable institutions and property rights, etc. (Solow 1970, Rosenberg and Birdzell 1986, Lucas 1988, Barro and Sala-i-Martin 1995) – is quite different from making detailed predictions about specific economic variables such as the rate of growth, employment, the price level, or the relationship of any particular sector or industry (such as energy) to the economy as a whole. And when it comes to long-term forecasting, as distinct from acquiring *insight* regarding long-term economic processes, the fact is that there simply are no economic models that have a track record of success.

Some forecasting exercises are held in deservedly low regard because of their pessimistic claims regarding resource exhaustion. A few environmentalists and others who have been concerned with global resource issues have contributed their share of exaggerations and inaccurate forecasts.[1] For example, the Club of Rome's notorious warning, *The Limits to Growth* (Meadows et al. 1972), gave wide publicity to a set of estimates made by the US Bureau of Mines projecting the "Static Index" (in years) of a variety of "nonrenewable natural resources." The Static Index was defined as "the number of years known global reserves will last at current global consumption" (US Bureau of Mines 1970). It was perhaps comforting to project in 1970 that there were sufficient reserves of coal to last 2300 years, and that reserves of aluminum, chromium, cobalt, iron, and the platinum group all exceeded 100 years. However, the Static Index values of the years of remaining reserves for gold (11 years), lead (26 years), mercury (13 years), silver (16 years), tin (17 years), and zinc (23 years) were such that, if the forecasts had been accurate, the world would be using exclusively recycled quantities of these metals by now. The Static Index of petroleum was 31 years, so that the 1970 forecast of the date at which we would have "run out of oil" was 2001.

Furthermore, there is a disconnection between the way the policy debate on climate change is dominated by economic models and forecasts, and the relative lack of influence of such techniques in business management. Few companies today maintain substantial economic forecasting units, and a wide spectrum of general business and economic forecasts can be obtained at relatively low cost. For example, a 12-month subscription to *Blue Chip Economic Indicators*, which contains the forecasts of more than 50 leading business economists (or their companies) of economic growth, inflation, interest rates, and other important economic variables, could recently be purchased for $627 per year (Moore 2002).[2] Although most of the information contained in *Blue*

Chip Economic Indicators pertains to the short run (one to two years), longer-run forecasts (for five years annually and then an average for five additional years) are also presented. These longer-run projections come with a disclaimer, however: "APPLY THESE TREND PROJECTIONS CAUTIOUSLY. The vast majority of economic and political forces cannot be evaluated over such a long time span" (Moore 2002, p. 14, capitalization in the original).

The poor track record of energy forecasting is hardly a secret. In a recent comprehensive survey paper on the "perils and promise of long-term forecasts," Koomey et al. (2001) provide numerous examples of forecasts that failed to be realized, from Jevons' gloomy nineteenth-century assessment of the future of England's energy system (which was coal-based at the time Jevons wrote) to long-term forecasts of US energy demand made during the 1970s (actual consumption by 2000 has turned out to be at the very lowest end of the range of the forecasts). Their advice to forecasters is "be modest; the future will probably unfold in ways that you haven't dreamed" (p. 2). A recent issue of *The Economist* (November 3, 2001), in an article describing controversy over the "Hubbert curve" (the pattern of oil production and decline originally put forward by Shell geologist M. King Hubbert) asked,

> Who is right? In making sense of these wildly opposing views, it is useful to look back at the pitiful history of oil forecasting. Doomsters have been predicting dry wells since the 1970s, but so far the oil is still gushing. Nearly all the predictions for 2000 made after the 1970s oil shocks were far too pessimistic. America's Department of Energy thought that oil would reach $150 a barrel (at 2000 prices); even Exxon predicted a price of $100. (p. 81)

This kind of skepticism is a reflection of the ineffectiveness of economic forecasting models, but there is a deeper reason at work as well. Specific, detailed knowledge about future economic trends or fluctuations could be converted into enormous wealth. For example, the ability to predict movements in the stock market on a daily, weekly, monthly, or yearly basis would enable the possessor of such information to amass huge gains from speculation. The existence of such potential gains would draw entrants, who would compete against each other in exploiting the sources of information about the future. The market positions taken by the agents with knowledge of the future would cause the trends that were the source of the speculative profits to disappear, and the extraordinary gains from speculation would be reduced to normal levels of profit from arbitrage and market research. This is the essential insight of the "efficient markets" idea that has come to dominate the economic theory of financial markets. The conclusion is that most fluctuations in economic variables *cannot in principle* be predicted. What is true about stock prices is true also about commodity prices – oil and natural gas no less than gold or pork bellies. The general proposition is that market competition tends to squeeze out opportunities for speculative profits, and hence that reliable knowledge of future economic variables is very difficult if not impossible to come by.

It is important to realize that the policy relevance of energy-economic models depends on their ability to project both prices and quantities of the various forms of energy used by the economy. The quantities of the various kinds of primary energy used determine GHG emissions, and the prices of energy are important in determining economy-wide effects. The models typically specify the impact of policy instruments (carbon taxes or energy taxes, cap-and-trade permit systems, etc.) through speci-fication of price changes, so the sensitivity of the models to energy price changes is a key component of the analysis. It is well-known that the "baseline" forecasts for prices and quantities are both the starting point and a major determinant of the results of any analysis of these market-oriented policies (Weyant 2000). This chapter will explore how suc-cessful economic-energy models have been in predicting energy prices, quantities, or both.

5.2.1 History of economic forecasts using energy-economy models

Because most GHG emissions arise from the use of fossil fuels to produce energy services, economic forecasting for climate policy has to start with the energy sector. Forecasting of energy demand and usage became an academic subspecialty during the oil price shocks of the 1970s. The impact of those price disruptions spawned a great deal of interest in modeling that was designed to predict the energy future of the United States and the rest of the world. In looking at the record of the models that were built during the 1970s to forecast the course of energy demand and prices, one finds that the forecasting performance of energy/economic models has not been accurate enough to justify their use in making policy judgments. Indeed, judging from their history, *the range of uncertainty in the forecasts of the best models typically is greater than the changes in energy demand that are the subject of the current climate policy debate.*

It should be noted that in order to do a retrospective analysis of the

performance of forecasting models it is necessary to use published forecasts that are old enough to make the forecasting exercise interesting.
Simply predicting that next year will look much like this year is unlikely
to be wrong by more than a few percentage points; but the time scale
relevant to climate modeling has to span a number of decades at least.
Forecasting techniques may improve over time, so the retrospective
analysis cannot ascertain the predictive power of *current* models. This
does not mean that the retrospective approach sheds no light on the
performance of current models, however. In some cases (such as the
National Energy Modeling System (NEMS) used by the Energy Information Agency of the US Department of Energy to construct the Annual
Energy Outlook (AEO) forecasts), the improvements have been in-
cremental rather than revolutionary. Notwithstanding the potential
advances in modeling technique, it is worth bearing in mind that
the models examined below were considered to be "state of the art" for
their time. The record of forecasting models in general can at a
minimum provide grounds for caution in accepting the claims of current forecasts.

Academic and government studies of the accuracy of forecasts when
compared to the actual performance of the energy sector show that the
forecasts have a dismal record. It should be noted at the outset that this
is not meant to be in any way a criticism of the forecasters themselves;
indeed, these scholars and analysts are part of the best tradition of
social scientific research. Their results have been published in the peer-
reviewed or monographic literature, or in reports that are available for
scientific review. The point is that even when the best representatives
of the modeling community are examined regarding the accuracy of
their forecasts, there is very little justification for optimism about long-
term economic forecasting ability.

An early comprehensive study of the reliability of forecasting is
William Ascher's *Forecasting: an Appraisal for Policy-Makers and Planners*
(1978). Ascher examined the accuracy of predictions in a number of
areas, including population, (macro)economics, energy, transportation,
and technological forecasting. In one set of comparisons, Ascher looked
at the accuracy of energy forecasts with target dates up to 1970 and with
target date of 1975 for three categories: total energy demand, electricity generated, and petroleum consumption. His results are interesting
in part because the pre-1970 target date forecasts were made in the
period before the first of the oil price shocks while the target date of
1975 occurred after this shock. Ascher found (not surprisingly) that for

the target dates of 1970 and earlier, the forecasts became more accurate the shorter the forecast period. Nevertheless, the errors were still large. For electricity consumption, the median error of 15-year forecasts was about 13 percent, compared to 6 percent for 10-year forecasts and 3.5 percent for 5-year forecasts. The 15, 10, and 5-year median forecast errors for petroleum were as large or larger: 13, 10, and 6 percent respectively. Not enough total energy forecasts were made with a pre-1975 target date to allow meaningful comparisons (Ascher 1978, 101ff.).

Even more instructive is Ascher's collection of 5- and 10-year forecasts having target years of 1980 and 1985, for each of the three energy categories. Although Ascher found "no indication yet of improved methodology in forecasting energy demand," he did state that "the only glimmer of hope for better forecasts in the future is the performance of the . . . econometric energy models" (1978, pp. 125–6). (His optimism may have been misplaced, as will be seen below.) Table 5.1 reproduces the forecasts in three energy categories, along with the actual values of each of the energy quantities and the percentage error for each of the forecasts. It should be noted that the forecasts reported by Ascher are in some cases median or average forecasts from the studies cited, so that the actual range of forecast errors would be greater than that computed in Table 5.1. The average absolute percentage errors for 1985 range from 30 percent (for petroleum consumption) to 48 percent (for total energy demand). The 1980 average absolute percentage errors are smaller, but still substantial.

A significant improvement in the sophistication and theoretical grounding of long-term energy forecasts is represented by William Nordhaus' (1979) monograph, *The Efficient Use of Energy Resources*. This study is notable because it very definitely represented the "state of the art" of energy modeling at its time. The model design and implementation are informed by economic theory, carefully estimated elasticities of demand for energy in different sectors, and the best engineering estimates of costs and supply availability. The models are unified in an optimizing programming framework, and the forecasts were constructed using the most sophisticated numerical methods that were available in the 1970s. In short, *The Efficient Use of Energy Resources* exemplifies the highest standards of applied economic practice at the time.

Even so, Nordhaus' longer-term forecasts of energy prices and consumption missed the mark by a considerable margin. One of the chief concerns of his monograph was appraisal of the extent of OPEC's monopoly power in the world oil market, so Nordhaus calculated three kinds of oil price forecasts: the "efficient" price path corresponding to

Table 5.1 Actual energy usage compared to 1974–76 forecasts (Ascher 1978)

Source	Total energy demand (10^15 BTU/yr)				Electricity generated by utilities (10^9 kWhr/yr)				Petroleum consumption (10^6 bbl/yr)			
	1980	% error	1985	% error	1980	% error	1985	% error	1980	% error	1985	% error
Actual value	76[a]	–	74[a]	–	2286	–	2470	–	6225	–	5740	–
Shell Oil	83	9	98	32	–	–	–	36	8300	33	8030	40
FEA[b]	–	–	99	34	2574	13	3348	62	–	–	7556	32
FEA[c]	87	14	102	38	–	–	3990	60	6077	–2	7423	29
Bureau of Mines	87	14	104	41	2769	21	3960	37	7433	19	8375	46
Data Resources, Inc.	90	18	106	43	2589	13	3383	–	7099	14	8158	42
Ford Foundation Energy Policy Project	92	21	107	45	–	–	–	46	6789	9	5475	–5
John Gray/NUS, Inc.[b]	–	–	113	53	–	–	3604	70	–	–	–	–
OECD[c]	97	28	114	54	3042	33	4207	–	6855	10	7556	32
AEC	–	–	118	59	–	–	–	–	–	–	–	–
John McKetta/University of Texas	107	41	138	86	–	–	–	30	–	–	–	–
Westinghouse	–	–	–	–	2516	10	3211	30	–	–	–	–
Joskow and Baughman/MIT[b]	–	–	–	–	2514	10	3217	31	–	–	–	–
Oak Ridge National Laboratory	–	–	–	–	2530	11	3245	–	–	–	–	–

Table 5.1 Continued

Source	Total energy demand (10^{15} BTU/yr)				Electricity generated by utilities (10^9 kWhr/yr)				Petroleum consumption (10^6 bbl/yr)			
	1980	% error	1985	% error	1980	% error	1985	% error	1980	% error	1985	% error
Livermore Laboratory	–	–	–	–	2603	14	3419	38	–	–	–	–
FPC Technical Adv. Comm.	–	–	–	–	2663	16	3564	44	–	–	–	–
A. D. Little, Inc.	–	–	–	–	2724	19	3715	50	–	–	–	–
ERDA[c]	–	–	–	–	3071	34	3890	57	–	–	7478	30
Young/University of California[b]	–	–	–	–	–	–	–	–	6130	–2	6480	13
Bradshaw/ARCO	–	–	–	–	–	–	–	–	–	–	6680	16
Bureau of Mines	–	–	–	–	–	–	–	–	–	–	6989	22
Independent Petroleum Association	–	–	–	–	–	–	–	–	–	–	8942	56
Average absolute error (%)		21		48		18		45		13		30

[a] Consumption.
[b] Median projection or scenario.
[c] Average of median projections or scenarios.
Sources: For the forecasts, Ascher (1978, Table 5.1, sources listed there); for the actual values, US EIA (1994).

perfect competition in all markets, various "limit price" paths corresponding to OPEC's possessing monopolistic power, and the "market" path corresponding to prices in a "realistic market environment." The "efficiency" and "limit" prices were intended to be bounding forecasts of the petroleum price, but the upper bound is fuzzy because the value of the limit price depends on what assumption is made about the relationship between the short-run and long-run price elasticity of oil. In practice, Nordhaus' "market" price forecast is higher than the standard-run "limit" forecast in the early years of the forecast period.

The reported model runs forecast prices out as far as 2050, but the most relevant comparisons are to the prices that have actually been observed as of 1975, 1985, and 1995. To facilitate comparisons and to put the prices in contemporary terms, the actual and forecast prices are converted to 1995 prices from the 1975 prices used by Nordhaus, using the chain-type GDP price deflator reported in a recent *Economic Report of the President* (Council of Economic Advisers 1998a). The comparisons are shown in Table 5.2.

This table has several features worth noting. In each of the three years shown, one of the Nordhaus forecasts is close to the actual price, but it is a different forecast in each year. The fact that the "limit" price is close to the actual price in 1975 is one of the reasons underlying Nordhaus' conclusion that OPEC had substantial market power during the period of the first oil shock, but in 1985 it is the "market" price that is close to the actual value, while in 1995 the "efficiency" or perfectly competitive price is close to the actual value. This pattern is consistent with a story that the world oil market became increasingly competitive after

Table 5.2 Actual and forecast crude oil prices, 1995 dollars/bbl (Nordhaus 1979)

Year	"Efficiency" price[a]	"Market" price[b]	"Limit" price (standard run)[c]	Actual price[d]
1975	7.78	33.10	23.04	26.57
1985	11.27	32.72	24.63	36.73
1995	17.97	–	37.76	17.22

[a] Nordhaus (1979, Table 6.5).
[b] Nordhaus (1979, Table 7.7).
[c] Nordhaus (1979, Table 6.5, Col. 10).
[d] US EIA (1998a), data on composite crude oil refiner acquisition costs. This cost figure corresponds to Nordhaus' crude oil delivered price.
[abcd] Prices converted to 1995 prices using chain-weighted GDP deflator from Council of Economic Advisers (1998a).

OPEC's heyday, but that would mean that a *political* forecast of the strength and cohesiveness of the OPEC countries over time would have been crucial to forecasting the actual path of oil prices over the decade and a half following publication of *The Efficient Use of Energy Resources*. The book itself contains no such political analysis.

There is a second difficulty with these forecasts. The Nordhaus model is an integrated model that forecasts both prices and quantities of energy. The lower the price that is forecast, the higher is the forecast quantity consumed. In the case of the Nordhaus model, the actual consumption of energy is lower than any of the forecast values for the post-1979 years. This is illustrated in Table 5.3. As can be seen from this table, the "market" projection for 1975 is close to actual consumption, but in this case the model's price is 25 percent too high. In 1985, the model errors in forecasting total quantity are 17 percent for the "market" forecast and 37 percent for the "efficiency" forecast; in 1995 these errors are 7 and 29 percent, respectively. Of course, what happened is that the Nordhaus model (and most other forecasts) did not anticipate the unprecedented increase in output per unit of energy that took place during the late 1970s and 1980s following the oil shocks. This is so even though the Nordhaus model carefully estimated price elasticities for the various energy-using sectors based on historical data.

Other energy projections of the 1970s and 1980s overestimated both the growth in consumption of energy in the United States and the future level of energy prices. In an influential article in *Foreign Affairs*, Amory Lovins compared forecasts of US energy futures along a "hard path" (a composite of projections by the Energy Research and Development Administration (ERDA), the Federal Energy Administration (FEA), the Department of the Interior, Exxon, and the Edison Electric Institute) and a "soft path" based on conservation and technological improvements (Lovins 1976). The "hard path" forecast of US energy

Table 5.3 Actual and forecast total energy consumption, United States, quadrillion BTUs (Nordhaus 1979)

Year	"Efficient" forecast[a]	"Market" forecast[a]	Actual consumption[b]
1975	82.5	71.9	70.6
1985	101.4	86.6	74.0
1995	117.6	97.0	90.9

[a] Nordhaus (1979, Table 7.5).
[b] US EIA (1997).

consumption in 2000 was in excess of 150 quads, while the "soft path" forecast was about 95 quads. Between the time the forecasts were made in the mid 1970s and the late 1990s, actual US energy consumption was less than *either* forecast,[3] and by 2000 the "hard path" forecast was more than 50 percent greater than the actual level of energy consumption.

Lovins' 1976 article generated a great deal of controversy when it was published. Congressional hearings were held; the hearing record (including submissions of Lovins and his critics) ran to over 2000 pages. This mass of testimony was excerpted and reprinted in "debate" format by Friends of the Earth (Nash 1979). One of the most extensive critiques of Lovins was offered by Harry Perry and Sally Streiter, two economists from National Economic Research Associates, Inc., an economics consulting firm. The Perry and Streiter critique (reproduced in Nash 1979)[4] contains a "Comparison of Energy Consumption Projections for the United States" compiled from seven major forecasting efforts. The projections go to 2025 in one case and to 2010 in two others, but six of the seven forecasts give an estimate of US energy consumption in 2000. Those projections are presented in Table 5.4. In some instances,

Table 5.4 Energy consumption projections for the United States, NERA compilation (publication dates 1974–77)

Study/case	2000 Projection (quads)	% error
2000 Actual	99.29	0
RFF Base Case	114.2	15
DRI-Brookhaven Base Case	156.2	57
DRI-Brookhaven Energy Tax Case	117.9	19
Energy Policy Project Historical Growth Case	186.7	88
Energy Policy Project Technical Fix Case	124.0	25
Department of Interior	163.4	65
Institute for Energy Analysis Low Case	101.4	2
Institute for Energy Analysis High Case	125.9	27
ERDA-48 Historical Base Case	165.5	67
ERDA-48 Improved End-Use Efficiency Case	122.5	23
ERDA-48 Coal and Shale Synthetics Case	165.4	67
ERDA-48 Intensive Electrification Case	161.2	62
ERDA-48 Limited Nuclear Case	158.0	59
ERDA-48 Combination Case	137.0	38
Average of forecasts or errors	142.8	**44**

Sources: Table 1 of Perry and Streiter (from Nash 1979, p. 354); US EIA (2000a).

more than one scenario is presented for a particular forecasting group. Table 5.4 lists projections of "gross" consumption, that is, consumption without omission of the losses that occur in electricity generation. Every one of the year-2000 projections overestimated actual consumption in 2000. This is the case even when scenarios were constructed with energy taxes or policy-induced efficiency gains.

Subsequent long-term forecasts have suffered from the same lack of predictive power. Consider the forecasts of energy prices made by the US Department of Energy in an official government study titled *Energy Security: a Report to the President of the United States* that was published in 1987. This report grew out of a broad-based interagency analysis and review conducted over late 1986 and 1987 to investigate the economic and security implications of oil imports to the United States.[5] According to this report,

> Outright prophecy about this country's and the world's energy future is too shaky a base on which to try to construct (or even evaluate) energy policy. For that reason, in conducting this study, the Department of Energy did not content itself with flat, simplistic predictions. Instead, DOE worked with a number of Federal Government agencies to develop several alternative energy-market scenarios through 1995. The two most important scenarios . . . present the upper and lower boundaries in an admitted range of uncertainty, thus offering at least some generalized projections about future US oil imports and OPEC production that might be considered reasonable.
>
> (US Department of Energy 1987, p. 20)

The world oil price (in 1985 dollars) of the "lower oil price case" was projected to be $22/bbl by 1995, and in the "higher oil price case" it was projected to be $28/bbl in 1995. If these are converted to 1997 prices (Council of Economic Advisers 1998a), they are equivalent to $32/bbl for the low-price scenario and $40/bbl for the high-price scenario. These are *upper and lower bound forecasts* (as of 1987), yet the price of crude oil in 1995 was actually only about $18/bbl (composite crude oil refiner acquisition cost, US EIA 1998a). The error in the 1987 price forecasts is thus between 78 and 122 percent, even though the forecasts were made only a decade earlier.

A recent retrospective analysis of five major energy studies that were conducted in the early 1980s reveals a similar pattern of overestimation of the forecasted energy price (Sanstad et al. 2001b). Sanstad and his coauthors reviewed energy projections by the US Department of Energy

Table 5.5 Projections of primary and delivered fuel prices and errors: five major energy studies of 1982–83 (all prices except world oil in 1996 dollars per million BTU)

Fuel	1982 Price	Median projection of 2000 price	Actual 2000 price	Median % error from actual
Primary				
World oil ($/bbl)	50.70	77.22	26.01	197
Wellhead gas	3.56	9.95	3.10	221
Minemouth coal	1.99	2.92	0.73	296
Delivered				
Natural gas (residential/commercial)	7.55	12.91	5.31	143
Electricity (residential/commercial)	30.35	32.09	20.55	56
Natural gas (industrial)	5.43	11.88	2.80	324
Electricity (industrial)	21.90	28.52	12.66	125
Residual fuel oil (industrial)	7.40	12.14	3.21	278
Gasoline (transport)	15.46	23.25	10.19	128

Source: Sanstad et al. (2001b, Tables 3A, 3B).

(1983),[6] the American Gas Association (1983), the Gas Research Institute (1982), Data Resources, Inc. (1983), and Applied Energy Services, Inc. (1983). All five studies slightly underestimated total US energy demand in the year 2000, but substantially overestimated energy prices. The median percentage error in the forecast of aggregate energy demand across the five studies was –4.4 percent. Table 5.5 shows the median percentage error in the price forecast for primary and delivered year-2000 energy prices. All but one of the forecast errors are greater than 100 percent, and four of the nine are greater than 200 percent. The analytical focus of Sanstad et al. is on how underestimation of the rate of increase in aggregate energy efficiency can skew the appraisal of policy options, but the data in Table 5.5 clearly indicates the lack of predictive capability of the five studies they analyzed.

5.2.2 Forecasting performance of the NEMS model

Perhaps the most prominent and influential energy/economic forecasting model in the United States is NEMS (National Energy Modeling

System), maintained by the Energy Information Administration (EIA) of the US Department of Energy. The NEMS model is used to construct the Annual Energy Outlook, published each year by the Department of Energy. The EIA also responds to specific requests by other parts of the government, as when it published in 1998 an analysis of the "costs of Kyoto" upon the request of the Committee on Science of the US House of Representatives, or its analysis of the economic effects of the regulation of pollution from US power plants requested by the Subcommittee on National Economic Growth, Natural Resources, and Regulatory Affairs of the US House of Representatives Committee on Government Reform (US EIA 1998b, 2000b, 2001a; see also 2001b).

The EIA also reviews the accuracy of its own forecasts (e.g., Holte 2000). Holte shows the average absolute percentage errors for five recent forecast evaluations (conducted in 1996, 1997, 1998, 1999, and 2000), but Table 5.6 reproduces only the results for the year 2000 forecast evaluation. The errors reported in Table 5.6 are defined as follows: "The

Table 5.6 Average absolute percentage errors from 2000 AEO forecast evaluation: AEO 1982 to AEO 2000

Variable	Average absolute percentage error
Consumption	
Total energy consumption	1.8
Total petroleum consumption	2.9
Total natural gas consumption	5.6
Total coal consumption	3.3
Total electricity sales	2.0
Production	
Crude oil production	4.5
Natural gas production	4.6
Coal production	3.5
Imports and exports	
Net petroleum imports	8.4
Net natural gas imports	15.9
Net coal exports	31.9
Prices	
World oil prices	55.7
Natural gas wellhead prices	68.2
Coal prices to electric utilities	36.6
Average electricity prices	11.8

Source: Holte (2000, Table 1).

average absolute forecast error is computed as the mean, or average, of all the absolute values of the percentage errors, expressed as percentage differences of the reference case projection from the actual, shown for each AEO, for each year in the forecast, for a given variable" (Holte 2000, p. 2). As in the Sanstad et al. study, the quantity forecasts have been more accurate than the price forecasts. Keep in mind that these averages cover forecasts made over the entire period from 1982 through 2000, and include both short-term (one year ahead, two years ahead, etc.) and longer-term forecasts. Not all of the AEOs contained forecasts reaching as far as 1999 or 2000; the AEOs of 1982–85 did not include price forecasts beyond 1995, for example. In addition, the later AEOs' longer-term forecasts cannot be compared to actual prices and quantities, because the actual values after 2000 (the year the analysis was published) could not be known. Thus, these average percentage errors are weighted towards shorter-term forecasts, which are likely to be more accurate than longer-term forecasts. Nevertheless, very large average errors in the price projections are evident.

The data presented by Holte also shows that even in the short run, the AEO forecasts contain little or no information about the actual future course of energy prices. If the forecasts had any predictive value, a regression of the actual prices as reported by the AEO on the forecasts should show the forecasts to have some explanatory power. In other words, if $p^f_{t,t-i}$ is the forecast of the price of oil (or natural gas) in year t that was made in year $t - i$, then a regression of p^a_t, the actual price in year t, on $p^f_{t,t-i}$ should exhibit a significant correlation. Yet for both oil and natural gas, even the forecasts made *only one year ahead* have no explanatory power. This is demonstrated by the regression results in Table 5.7. Neither regression provides any evidence for rejecting the null hypothesis that the forecasts have no predictive power even for the price only a year ahead. These results are consistent with the market efficiency notion that the publicly available forecasts of the EIA can offer traders no insight into future market developments.

A different methodology to evaluate the accuracy of the EIA's forecasts was employed by Shlyakhter et al. (1994). They examined two types of energy demand forecasts: a cross-section of 69 energy demand forecasts projected for the year 2000 and projections for approximately 180 energy producing or consuming sectors contained in the EIA's Annual Energy Outlook.[7] Shlyakhter and his coauthors were interested in comparing the magnitude of actual forecast errors to the reported uncertainty in the forecasts as measured by the implicit "standard deviations" of the forecasts. However, most published forecasts, when they

Table 5.7   Estimated regression equation, $p_t^a = a_0 + a_1 p_{t,t-1}^f + \varepsilon_t$, 1985–99, world oil price and natural gas wellhead price from Annual Energy Outlook

	$\hat{a}_0$ (Standard error)	$\hat{a}_1$ (Standard error)	R^2	Adjusted R^2	Regression Probability-Value
Oil	12.241 (5.746)	0.271 (0.278)	0.080	−0.004	0.349
Natural gas	1.264 (0.607)	0.319 (0.284)	0.103	0.021	0.286

Source: See text. Data from Holte (2000). Both actual and forecast price variables are in nominal terms (dollars per barrel for oil and dollars per thousand cubic feet for gas). There was no AEO 1988 and AEO 1990 did not have a forecast for 1991, so each regression was based on 13 observations.

report "error bounds" at all, do not express them in terms of a statistically derived standard deviation, because the error bounds arise primarily from uncertainties in the assumptions driving the forecasts (such as the rate of future economic growth) rather than uncertainties in the statistical estimates of model parameters (Shlyakhter et al. 1994, pp. 119–20). To overcome this obstacle, Shlyakhter et al. convert reported upper and lower bounds into an estimate of the standard deviation of the forecast by making an assumption about the underlying distribution characterizing the uncertainty of the forecasts. That is, they assume that the difference between the upper (U) and lower (L) bounds is k times the implicit standard deviation of the distribution of forecasts (represented by Δ), or

$$k\Delta = (U - L) \tag{5.1}$$

They then define the "forecast error ratio" x as

$$x = (T - R)/\Delta = k(T - R)/(U - L) \tag{5.2}$$

where T is the true (realized) value of the variable being forecast and R is the forecast (the mean or "reference case" of the forecast distribution). If the underlying distribution of the forecasts is normal, it is possible to predict the frequency by which $|x|$ exceeds any particular value. If $k = 2$ (the value assumed by Shlyakhter et al.), then $\Pr(|x| > 1) = 0.32$, $\Pr(|x| > 2) = 0.045$, etc. (If $k > 2$, these probabilities would be correspondingly smaller.)

Comparison of the actual outcomes to the "reference case" forecasts reveals that the likelihood of large forecast errors is much greater than would be predicted if the underlying distribution of forecasts were in fact normal. Thus, for the AEO forecasts, the actual likelihood that the forecast error ratio is greater than 1 is approximately 0.7, and the likelihood that it is greater than 2 is 0.5.[8] The same magnitude of forecast errors (relative to the implicit standard deviation of the forecasts) was found for each set of projections – the implicit standard deviations fell as the forecast interval decreased, but the forecast error ratios did not. The performance of the forecasts did not improve when the individual sectors were aggregated, contrary to the authors' prior expectations (Shlyakhter et al. 1994, p. 123). Their survey of 69 energy forecasts for the year 2000 (a sample distinct from the AEO sectoral sample) showed somewhat lower forecast error ratios, with $\Pr(|x| > 1) = 0.6$ and $\Pr(|x| > 2) = 0.2$. These are still substantially larger than if the underlying distribution were normal. The likelihood that individual forecasts fall outside a particular distance from the mean is smaller in this case because of the much greater diversity of the forecasts (which thereby produce a larger implicit standard deviation).

What these results mean is that the actual uncertainty underlying the energy demand forecasts reviewed by Shlyakhter et al. is much larger than suggested by the forecasts' "upper" and "lower" bounds. The retrospective analysis by Shlyakhter et al. suggests that the true values will be farther from the reference case than the upper or lower bound of the AEO forecast about 70 percent of the time.

5.2.3 Calibration errors in energy-economic models

There certainly is no reason to be confident about the forecasting performance of conventional energy/economic models, based on their record. There is another reason, not commonly recognized, to be skeptical of the models as they are presently constituted. The fuels that make up the heart of the energy sector – coal, oil, and natural gas – are commodities having large stocks relative to their annual consumption. As such, they are "storable" as opposed to "non-storable" commodities (such as perishable fresh fruits or vegetables).[9] Williams and Wright (1991) have shown that the array of present and future prices of storable commodities is linked by the economic costs and opportunities for storage, and that any models that fail to take these linkages into account are bound to give inaccurate and misleading results. Yet the intimate connection between spot and future prices is absent from the major forecasting models presently in use.

For example, consider the estimate of the "costs of Kyoto" that was produced by the EIA using its NEMS model (US EIA 1998b). The model operates in such a way that "[i]n the end-use demand sectors, foresight is assumed not to have a material influence on energy equipment decisions, and such decisions are modeled on the basis of prices in effect at the time of the decision" (US EIA 1998b, p. 20, fn. 22; see also p. 7). Only in the refining and power generation sectors are the producers allowed to exercise foresight, and in the generation sector this foresight is assumed to be perfect in the sense that "[a]n algorithm solves for the path of carbon prices in which anticipated and realized carbon [permit] prices are approximately the same" (US EIA 1998b, p. 20, fn. 22).

Nevertheless, futures prices are connected to current spot prices and market conditions (the level of inventories, anticipated weather conditions, etc.); indeed, the spot and futures prices constitute an array of interrelated prices (Williams 1986) that are linked by arbitrage to transportation costs, storage costs, and interest charges. Models such as NEMS are calibrated to yield current prices as the outcome of their market-clearing conditions, but could in principle also be calibrated to incorporate the prices of the various futures contracts that are traded in the market. If the structure of the models were "correct" (that is, if they adequately reflected all market and behavioral conditions), the price forecasts of the models could be accurately linked to futures prices prevailing in the current market. After all, these futures prices are tangible prices just as surely as the spot price of the commodities being traded; ownership of a futures contract gives the holder the legally enforceable right to receive delivery of a specified quantity of a particular commodity (standardized with respect to grade and quality) at a particular time and a particular location. The price of such a futures contract is just as real as the spot price prevailing at any given moment in time.[10]

Given that future spot prices, current spot prices, and futures prices are linked, a properly specified model would incorporate all the information contained in spot and futures prices in its projections of the evolution of commodity prices over time. However, according to the Williams/Wright analysis, the variance in futures prices declines as they pertain to more distant dates – asymptotically no additional information is contained in a more distant futures price than what is already included in the existing prices. The basic point is that only a limited amount of information about the future is available to the present. Nothing else is accessible. Even in stylized models in which the structure of the markets, costs of storage, and nature of the intrinsic randomness of the economy are explicitly specified, the futures prices can predict realized prices only with R^2 values between 0.1 and 0.3 (Williams and Wright 1991, pp. 176 ff.).

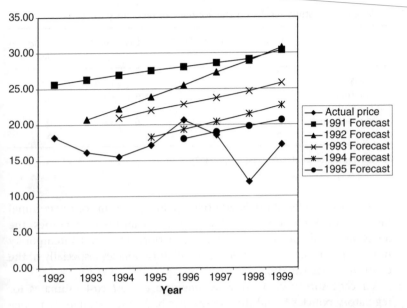

Figure 5.1 Comparison of actual and forecast oil prices, 1992–99

Market values of oil and natural gas futures contracts are readily available. Nevertheless, none of this information is incorporated into NEMS or any of the other models used in climate policy analysis. In NEMS, the price forecasting model appears to be some kind of simple rising trend, recalibrated every year to the current level of prices. Figure 5.1 shows a sample of the actual and predicted prices for oil taken from the successive Annual Energy Outlooks and reported in the Holte study (2000). It is evident that in every case the forecast prices simply increase gradually over time. Of course, actual oil prices fluctuate up and down from year to year. The same pattern holds for natural gas prices as well.

5.3 Predictions of the costs of greenhouse gas reductions and other regulatory policies

Even if energy-economic models do a poor job in forecasting the course of the economy or its energy sector, it is still possible that the economic models presently in use are capable of providing policy guidance because they can be used to calculate the economic consequences of alternative policies. Economic modeling is used to estimate the impact

Table 5.8   Case study results on the accuracy of regulatory forecasts

	Accurate	Overestimate	Underestimate	Unable to determine
Quantity reduction	13	9	4	2
Unit pollution reduction cost	8	14	6	0
Total cost	5	15	3	5

Source: Harrington et al. (2000, Table 2, p. 307).

of changes in taxes, more liberal trade policies, reforms of the criminal justice system, etc. Instead of trying to cover all these literatures, the focus here will be on reviewing the performance of contemporary models in calculating the effects of regulatory policies (especially of the environmental variety).

A recent survey comparing *ex ante* and *ex post* cost estimates for regulatory policies found that the studies had a predominant tendency to overestimate the costs *ex ante* (Harrington et al. 2000). Harrington and his coauthors reviewed 28 cases in which *ex ante* and *ex post* cost estimates were calculated in studies not generated by the regulated industries themselves. A forecast was classified as "accurate" if the *ex post* (realized) result fell within the error bounds of the *ex ante* study or if it varied by no more than ±25 percent from the point estimate of the *ex ante* study. Three categories of forecasts were examined: the quantity of pollution reduction, unit pollution reduction cost, and total cost. Table 5.8 reproduces some of the results of their review. Two main features of the *ex ante/ex post* comparison are evident from this table. First, the forecasts are not at all accurate on the whole. Even with the very generous definition of "accuracy," fewer than 30 percent of the forecasts were accurate in predicting either the total cost or the unit cost of the regulations. Second, the tendency of the *ex ante* studies to overestimate regulatory costs can be seen. This finding is consistent with the earlier study by Goodstein (1997), which included industry *ex ante* estimates in the comparison set (estimates that Harrington et al. excluded).

There are several notorious examples of the overestimation of regulatory difficulty or cost in the environmental area.[11] *Ex ante* industry estimates of the per ton cost of SO_2 emissions reductions under the Clean Air Act Amendments of 1990 were as high as $1500 per ton; the actual cost of the emissions permits as of 2000 were about $200 per ton, an increase from \$75 per ton in 1997.[12] The forecast was off by a factor of 7.5 in this case. Less well known is the case of estimates of the cost of eliminating ozone-depleting substances under the Montreal Protocol. Hammitt (1997) found that the actual reductions in CFC-11 and CFC-12 were much greater (given the price increases following the signing of the Protocol) than had been predicted *ex ante*. Even more dramatic is that the RAND Corporation, in work carried out for the US EPA and published in the early 1980s, estimated first that "the most promising set of mandatory controls could reduce cumulative emissions [of CFCs] over the period [1980–90] by perhaps 15 percent" (Palmer et al. 1980) and then that the available technical options had the potential to reduce CFC emissions by only about one-third (Mooz et al. 1982). It was thought to be so difficult to find substitutes for CFC-113 as a solvent in electronics manufacture that there was discussion of excluding it from the Montreal Protocol altogether (Benedick 1991). In actuality, CFC production (except for a few quantitatively unimportant "essential uses") was phased out entirely in the developed countries by the end of 1996, and will be eliminated in developing countries by 2007. This has been achieved with no discernable effect on lifestyles or the standard of living, and was accomplished in some cases at a net gain to producers and consumers (Cook 1996).

There is no way yet of knowing the impact of mandatory GHG reductions on the US economy, because no regulations have been put in place that would bring about such reductions. However, it is possible to compare different model estimates of the GDP cost of such reductions. Just such a comparison has been carried out twice in the 1990s by the Stanford Energy Modeling Forum. The EMF, in 1993 and 1999, coordinated a set of runs, by the leading energy-economic modeling groups participating in the EMF, of comparable emissions reductions scenarios. In the 1993 (EMF-12) comparison, the GDP impact of two alternative carbon reduction scenarios was simulated, while in the 1999 (EMF-16) runs, the GDP effects of compliance with the emissions reductions targets of the Kyoto Protocol were calculated under alternative assumptions about the extent of international allowance trading. The results of these comparisons are summarized in Table 5.9. This table shows considerable variation in the results from the different models in both EMF exercises, whether measured as the ratio of the maximum estimate of the GDP change to the minimum estimate of the GDP change or as the coefficient of variation of the estimated GDP change.[13]

In both EMF-12 and EMF-16, the models were run under uniform scenario definitions and the results varied considerably. Yet some of the

Table 5.9 Range of estimates of 2010 GDP effects of GHG reduction policies, EMF-12 and EMF-16

Model	EMF-12: scenario (% 2010 GDP loss)		EMF-16: meeting the Kyoto target (% 2010 GDP loss)		
	Stabilization	20% reduction	No trading	Annex I trading	Global trading
CRTM	0.2	1.0			
DGEM	0.6	1.7			
ERM	0.4	1.1			
Fossil 2	0.2	1.4			
Global 2100	0.7	1.5			
Goulder	0.3	1.2			
GREEN	0.2	0.9			
MWC	0.5	1.1			
ABARE-GTEM			1.9	0.8	0.2
MS-MRT			1.9	0.9	0.3
CETA			1.8	0.6	0.4
MERGE3			0.9	0.5	0.2
RICE			0.9	0.6	0.2
AIM			0.4	0.3	0.2
G-Cubed			0.4	0.2	0.1
Maximum/ minimum	*3.5*	*1.9*	*4.8*	*4.5*	*4.0*
Coefficient of variation	*0.51*	*0.22*	*0.57*	*0.45*	*0.42*

Sources: For EMF-12, see IPCC (1996). For EMF-16, see Weyant and Hill (1999) and Sanstad et al. (2001a). In EMF-12, the stabilization scenario entails stabilization of CO_2 emissions at their 1990 levels by 2000; the 20% reduction scenario means stabilization at 1990 levels by 2000 and 20% reduction below 1990 levels by 2010. The EMF-16 estimated losses were converted to a percentage of GDP using the forecast of 2010 GDP in IWG (2000), converted to 1990 dollars using the GDP deflator from US Department of Commerce, Bureau of Economic Analysis (2001).

elements of the scenarios are in fact policy options that could be subject to change. All the EMF scenarios assumed that the tax revenues raised by a carbon tax sufficient to bring about the requisite CO_2 emissions reductions are returned to the economy as "lump-sum transfers" – as payments that have no effect on the incentives of producers or consumers. However, it is well-known that different methods of "revenue recycling," that is, of returning the tax revenues to the economy, have very different GDP effects (Repetto 2001, Sanstad et al. 2001a).

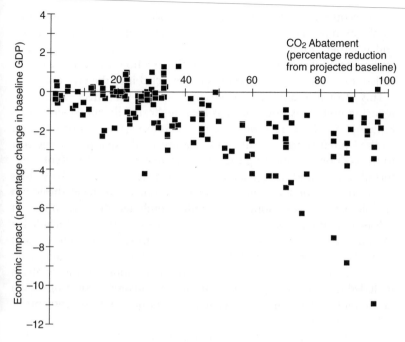

Figure 5.2 Variations in GDP impact: 162 simulations across 16 models. *Source*: Repetto and Austin (1997). Reproduced with permission

Returning the revenues in the form of reducing the marginal rates of other distortionary taxes (such as taxes on capital or on labor income) is more beneficial to the economy than lump-sum revenue recycling. Furthermore, the type of revenue recycling is not the only policy option that might be varied. Repetto and Austin (1997) have shown that even the *sign* of the GDP effect of the GHG reduction policies depends on which assumptions are made in the modeling exercise. Figure 5.2 shows the variation in GDP impact that results when different assumptions are made and different levels of CO_2 abatement are projected. Repetto and Austin found that eight assumptions (in addition to the size of the projected CO_2 reduction) account for 80 percent of the variation in predicted economic impacts. The eight assumptions specify the availability and cost of a noncarbon "backstop" technology, the efficiency of the economy's response to price changes, the degree of interfuel and product substitutability, how much time before the specified CO_2 reduction target is achieved, whether the CO_2 reduction would avoid some of the economic costs of climate change, whether reducing fossil fuel combustion would avoid other (nonclimate) air pollution damages, and whether "joint implementation" such as international emissions trading options are available, in addition to how the carbon tax or emissions permit auction revenues are recycled into the economy.

This strong finding that even the sign of the effect of carbon-reducing policies is uncertain has been replicated in subsequent work. Krause et al. (2002) reviewed five major studies generated or relied on by the United States government in addressing the question of the cost to the economy of meeting the emissions reductions of the Kyoto Protocol.[14] Krause and his collaborators found that while all of the studies included some of the available policy instruments to meet the Kyoto emissions reductions, none of the studies combined all these policies in a least-cost integrated framework. The potential policies included a carbon tax or permit auctions, market and institutional reforms and technology programs, tax shifts, international allowance trading, and enhancement of carbon sinks. In addition, none of the analyses included the full range of benefits from CO_2 emissions reductions, such as the monetized value of associated local air quality improvements.

The most pessimistic GDP impact estimated was a 4.2 percent loss in the EIA scenario involving only a domestic carbon tax and no enhancement of sinks. (The least pessimistic EIA scenario, which included global trading, sinks, and some tax shifting, showed a GDP loss of less than 1 percent.) On the other hand, the two "five lab" studies by the Interlaboratory Working Group showed slight GDP gains, primarily because the market reforms and technology policies emphasized in those studies moved the economy towards its production-possibilities frontier (though neither of the five-lab studies achieved the full Kyoto target by 2010). The benefit from a least-cost integrated approach to Kyoto is shown in Figure 5.3. Combining international permit trading, tax shifts, and the market, organizational, and institutional reforms outlined in the five-lab studies results in a net GDP gain of approximately $57 billion (1997 dollars) in 2010. The most important component of this gain is the efficiency enhancement representing movement towards the production-possibilities frontier (the market reforms).

These results demonstrate that the outcome of economic analyses of the impact of moderate GHG reductions in the United States depend primarily on *modeling assumptions* (including definition of the system boundaries, i.e., whether environmental benefits are included in the analysis). This robust result underscores the irony of the fact that while a great deal of attention has been focused on uncertainties in the

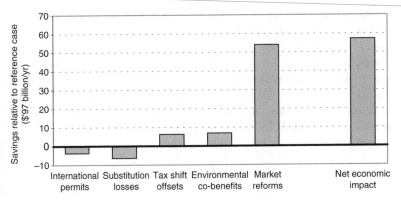

Figure 5.3 US cost and savings components of a least-cost Kyoto strategy with global trading, 2010. *Source*: Krause et al. (2001)

physical science models of the dynamics of global warming, estimates of the economic impact of various policy responses have been taken at face value. In fact, the uncertainties inherent in the economic analyses are much more serious – as noted above, the physical models are based on theoretically consistent and experimentally confirmed laws governing physical and chemical processes, while the economic models are grounded primarily in the assumptions made by the modelers.

5.4 Conclusions and implications for policy

It would be going too far to claim that long-run energy/economic models have no value just because they have little or no power to predict the course of energy prices and demand. Long-term modeling exercises can be valuable for providing a consistent framework in which to work out the consequences of various scenarios. This kind of analysis is not the same thing as prediction, and the point that should be emphasized is that its primary use should be to explore alternative *assumptions*.

At the same time, the poor predictive performance of models of this type indicates that considerable caution should be used in interpreting the results of model runs. If technological change is the main driver of economic change over the long run, it would be unwise to place too much faith in models that represent technological change in only the most stylized fashion (Sanstad 2000). If not only the magnitude, but even the sign of the effect of a proposed policy depends on contestable model assumptions, then the results of the analysis need to be taken with a very large grain of salt. Energy modeling in economics has a long enough history that the match between forecasts and actual outcomes can be compared. The result of such comparisons is not encouraging for those who would invest economics with an aura of scientific precision. This does not mean that economists should abandon the effort to gain insight into the forces shaping the long run; it only means that they must be conscious of the limits to their ability to do so.

6
Principles for the Future

The previous chapters have shown that by downplaying or ignoring the issues of *distribution, multiple equilibria,* and *dynamics,* the economic analysis of climate policy has missed (or mischaracterized) essential aspects of the problem. Most analysis has concentrated on matters of efficiency that are more suitable for small-scale environmental problems (such as where to locate a landfill or what to do about pollution of a river by an industrial plant) than for a global issue like climate change. By implicitly taking the existing definitions and distribution of property rights for granted, policy advice has been skewed towards the status quo. Some of the possibilities that are opened up by the creation of climate rights or alternative evolutionary paths for the economy have been ruled out arbitrarily, when they may actually offer the most hopeful potential routes to climate stabilization.

The result of this modeling failure has been a bias against bold and timely action, overestimation of the cost of emissions reductions, and pervasive paralysis in the political debate. By treating technology as a constraint on production, rather than as a dynamic process of innovation, the consequences of climate policy initiatives have been seen almost entirely in terms of adverse trade-offs. By ignoring the wealth effects associated with the creation of new climate rights (and from the destructive consequences of large-scale climatic disruptions), the welfare analysis of climate change has been couched almost exclusively in terms of changes in ordinary GDP (sometimes with climate damages "monetized" at current prices). This means that the "costs" of climate change have been estimated as marginal displacements of consumption or GDP. The strong negative impact of the *risk* of climate disruption has been systematically downplayed. No attention has been given to the potential constituencies who would favor climate policy activism.

Nothing has been more destructive of progress and consensus in the arena of international climate diplomacy than the focus on efficiency to the exclusion of equity concerns. The United States in particular, throughout the Kyoto negotiations and up to the very end of our meaningful participation when the Bush Administration repudiated the treaty, clung obsessively to the need for "flexibility mechanisms" such as international trading of emissions rights as a precondition for our involvement. The reason was a determination to lower the purported costs of the policies through purchase of Russian "hot air,"[1] taking advantage of low-cost emissions reduction opportunities in developing countries, and counting enhanced carbon sinks as part of the national emissions reductions. US government analyses consistently placed great store in the cost-reducing possibilities of international trading of emissions rights, even though more comprehensive calculations show that a determined domestic emissions reduction strategy could achieve the Kyoto target at no net loss to the economy (Krause et al. 2002).

Of course, attempting to negotiate the devilish details of the flexibility mechanisms created a nightmare of complexity. Questions of reporting, compliance, and enforcement loomed large. As the developed country representatives attempted to work out an acceptable set of rules, the developing countries remained largely outside the process. The developing countries were determined to address the *equity* issues involved in setting current and future emissions targets. They also wanted to guarantee the kinds of technology transfers and capital flows that would forestall any conflict between climate protection and their own economic growth.[2] While the United States, driven by its insistence on economic efficiency, demanded international permit trading regardless of what the emissions targets might be, the developing countries focused on how to rectify the large historical discrepancies in *cumulative* emissions resulting from the fact sustained economic growth had begun much earlier in the developed countries. "Meaningful participation by developing countries" in a Kyoto-like emissions reduction scheme was thus subject to conflicting interpretations: the rich countries rightly saw that without full global participation, any climate protection program was ultimately doomed to fail over the long run. In opposition, the developing countries feared institutionalization of something like current emissions levels (or ratios) that would condemn them to permanent economic inferiority because of the advantages the rich countries had derived from their historic reliance on fossil fuels to power the industrial revolution.

The preoccupation with efficiency by the United States was an example of the bias created by taking existing definitions of property rights for granted. Ignoring the wealth creation and destruction that must accompany any meaningful change in the pattern and cost of fossil fuel use (or perhaps more accurately, addressing the distributional issues only indirectly rather than openly and transparently), the US position sowed seeds of suspicion and distrust that plagued the negotiations from the start. A per capita based assignment of property rights in carbon emissions would have immediately created a powerful constituency for a climate protection regime in the developing countries, because their populations would have received the largest share of the carbon rights. If such an equity foundation had been laid, the rich countries would have faced cooperation rather than opposition from the developing world, and would have been free to move ahead in the design of efficient market mechanisms based on a distribution of carbon rights that would command support of the poorer countries. Such an assignment of climate rights would entail a substantial transfer of wealth from the developed to the developing world, but it is entirely possible that the citizens of the rich countries would be willing to accept such a transfer (and the accompanying changes in their energy use patterns) if they felt that the policies were actually going to be effective in solving the climate and development problems over the long run.[3]

Kyoto did not resolve these tensions. Although the emissions reductions called for by the treaty were significant (for the United States, 7 percent below 1990 levels as of 2008–12 would have amounted to roughly a 30 percent reduction in emissions from the business-as-usual baseline), they are far short of what will be required to avoid "dangerous anthropogenic interference with the climate system" (the phrase from the text of the 1992 United Nations Framework Convention on Climate Change).[4] Yet even Kyoto's first steps could not be achieved without significant policy initiatives including a carbon tax or emissions permit system. While the setting of emissions standards through international negotiations could be interpreted as a partial "loss of sovereignty," the new property rights implicitly created by such a process could be part of a larger consensus solution to the climate problem. Rather than seeking short-term national advantage in lawyerly fashion, the United States could have taken the lead to reach international agreement based on the creation of new climate rights.

It should have been acknowledged from the outset that the creation of these rights would have a significant (but not overwhelming) effect

on personal and national wealth. The redistribution would not necessarily have a permanent effect on any nation's or region's prosperity,
however. If there is sufficient intergenerational economic and social
mobility, the consequences of *any* initial distribution of property die
out over time through the normal working of the market process. The
voluntary market exchanges that take place in a dynamic economy
guarantee that the future pattern of wealth ownership is almost entirely
independent of the initial distribution (DeCanio 1992). The proper role
of the market is to decentralize decision-making as much as possible
given a suitable definition of property rights. When a new problem such
as anthropogenic climate change develops, it is appropriate to extend
the scope of property rights to bring the externality under control. Efficiency gains realized through markets then have an important role in
maintaining economic growth and vitality. However, it is a distortion
of the virtues of the market economy to argue that a status quo based
on an *incomplete* set of property rights is somehow "optimal" or that to
protect the environment necessarily involves economic losses.

It is also the case that creation of new climate rights would enable
the political process to "manage inequality" in a relatively benign
fashion. Redistribution can turn ugly, especially when currently defined
rights are taken away from some and given to others in a coercive zero-
sum fashion. In contrast, the creation of *new* rights offers the opportu-
nity to associate the inevitable redistributions with a higher social
purpose – in this case stabilization of the climate. The climate problem
has to be solved if humanity is to survive in the long run; it is also the
case that sustainable development and the elimination of mass poverty
are required if the future is to be stable and peaceful. Either of these
twin necessities would in itself justify a fresh look at property rights
assignments.[5] An intelligent policy of climate rights creation could
address both problems at once.[6] In any event, it is futile to adhere to
the fiction that doing nothing is the conservative course – inaction
redistributes rights as surely as conscious action, as the damages of
climate change and risks of climate catastrophe make themselves felt.

The preceding chapters have shown how assignment of climate rights
affects all aspects of the economic system, including interest rates. If we
are to avoid the moral arrogance of selecting the present as the preferred
temporal vantage point, it is also clear from the simple models worked
out earlier that some kind of ethical system balancing the benefits and
obligations of different generations is required. Neither the market
system nor democratic majoritarianism offers a ready-made solution.
Rules based on deeper notions of intergenerational justice are necessary.

The development of such principles is the subject of a different study, although it seems plausible that ideas such as the Golden Rule would occupy a central place in such an endeavor.[7]

In much the same way that equity and efficiency cannot be disentangled, so it is with economic dynamics and the progress of technology. We have seen that neither the modern theory of the firm, the evidence on intra-industry heterogeneity, nor the logic of evolutionary models supports the representation of production by the "profit maximization subject to a production function" model. Standard general equilibrium models do not embed a coherent dynamic story, so taking dynamics as the *starting point* for the modeling of production would result in no loss of information but would have the potential for significant gains in realism.

An important consequence of the results developed in the previous chapters is that claims to be able to estimate precisely the "costs" of climate protection policies must be abandoned. If firms are not on a well-defined production-possibilities frontier, and if the dynamics of technological progress are only imperfectly understood, then any hope of exact calculation of the costs of alternative policies vanishes. Unconventional policies, such as voluntary pollution-reduction programs, government–industry partnerships, and internally generated corporate "greening" campaigns can be effective. At the same time, the effects of a carbon emissions price increase – substitution in favor of less carbon-intensive products, the accelerated development of energy-efficiency technologies, and the promotion of market and institutional reforms that improve productivity in multiple dimensions – cannot easily be disentangled.

A corollary is that it is fruitless to attempt to determine the "optimal" carbon tax. If neither the costs nor the benefits can be known with any precision, just about the only thing that can be said with certainty about the welfare-maximizing price of carbon emissions is that it is greater than zero. Economists have a great deal to say about how to implement such a tax efficiently and effectively, about the similarities and differences between a tax and a system of tradable carbon emissions permits, and about the best way to recycle the revenue from such a tax or permit system. And, as we have seen above, the distributional consequences of such a tax or permit auction plan will affect other economic variables through system-wide feedbacks. However, any attempt to specify the exact level of the "optimal" tax is less an exercise in scientific calculation than a manifestation of the analyst's willingness to step beyond the limits of established economic knowledge.

Despite the limitations of their models, economists have been willing to play a central role in the policy discussion, and there is no doubt that they have had a great deal of influence on the debate. Power is seductive; susceptibility to its attractions is not evidence that economists as a group are more self-promoting or venal than natural scientists or other social scientists.[8] Instead, the failure of economics derives from a kind of *hubris* on the part of the profession. The technical achievements of economics, particularly of mathematical general equilibrium theory, have engendered an attitude of superiority vis-à-vis other disciplines. More seriously, the limitations and unanswered questions that lie at the heart of general equilibrium theory have been ignored in the drive to provide answers that can carry political weight.

In fairness, it is difficult to imagine what a true social *science* would look like if it were not based on some simple unifying principles. The danger, however, is in settling on abstractions that leave out essential features of the social process. By reducing human behavior to axiomatically defined rationality and maximization, it is possible to confuse real *economic* problems with purely *mathematical* ones. Models can always be set up in such a way as to be analytically tractable, whether or not mathematical convenience is a feature of the economic reality the models are intended to represent. The ability to solve mathematical puzzles, while requiring a great deal of ingenuity and high-powered intellectual skills, conveys a sense of power and accomplishment that is easily mistaken for *real wisdom*. Thus, economists can slip into the mode of offering advice and policy prescriptions on the basis of the results of their models, even though the models do not adequately capture the key features of the social world.

It is perhaps understandable that economists desire to emulate the natural scientists by seeking sparse and elegant mathematical descriptions of the social system. Unfortunately, this quest has so far proven to be futile. As discussed in the previous chapters, it has been known for some time within economics that the most parsimonious representations of rationality do not provide sufficient restrictions on the behavior of aggregates of diverse individuals to determine a unique equilibrium. There are limits on what economics can say, given only the requirements of formal rationality. By disregarding these limits in carrying out climate policy analysis based on neoclassical general equilibrium theory, economists have constructed an imposing structure on unsound foundations. While the edifice has the trappings and appearance of scientific rigor, its "results" are in fact derived in large part from unverified or untrue assumptions.

Despite its limitations, economics can make an important contribution to the formation and implementation of a climate policy that treats all people and all generations fairly. Just because general equilibrium models cannot provide the kind of precise guidance that politicians wish for, it does not follow that economics appropriately conceived is devoid of useful insights. Before Kyoto, at least, there was actually quite a high degree of consensus about climate policy among a wide range of economists. The Economists' Statement on Climate Change of early 1997 reflected this consensus – it called for action on the problem, international cooperation, and use of market mechanisms to implement regulatory action. It noted that responsible policies would not harm the US standard of living and might even "improve US productivity in the longer run."[9] Economists understand the power of compound growth, and realize that even if policies to reduce GHG emissions reductions were to reduce GDP by 1 percent below a business-as-usual baseline permanently, this would only cause a delay of approximately six months in achieving any given level of per capita income (based on the economy's historical annual real per capita growth rate of around 2 percent).

The advice offered by the Economists' Statement on Climate Change is quite broad and vague as to specifics. Practical policies have to be more explicit about just what carbon tax rate would be enacted, exactly what would be done with the revenues, how much and what kinds of R&D would be subsidized, and so forth. Economists have much to say about such issues. There should be nothing surprising about this – economics began as "political economy," after all, and part of its purpose always has been to inform policy. And because so much of what passes for "economic policy" generated by politicians and special interest groups is so stupid and misguided, economists have good reason to feel superior. (Think about rent controls, protectionist trade barriers, and poorly drawn regulations.)[10]

These considerations do not give economists carte blanche to say what should or should not be done to protect the global environment, however. Some of the models being used for climate policy work evolved from the energy/economic models developed in the 1970s in response to the oil price shocks. It is not at all clear that the framework appropriate for analyzing the earlier problem is well-suited for the current situation. The climate problem hinges on issues that are very poorly handled by conventional energy economics – matters of distributional equity, nonmarket environmental quality, intergenerational fairness, and the internal workings of firms. Instead of relying on flawed general equilibrium models to point the way, we could start from a specification of the characteristics of the future (or potential futures) we wish to bequeath to our successors and work backwards.[11] Economics has much to contribute to policies that can bring about a desirable future. From the arithmetic identity that points to the need for decarbonization (and more generally, dematerialization) of production[12] to the design of incentives that actually can bring about intended outcomes, economic knowledge and insight should be an integral part of the policy process.

At the most fundamental level, however, while economics can offer insight it does not provide solutions. An ethical foundation for climate policy must be established that recognizes the rights and obligations of all generations. This foundation has to recognize that not all affected parties are able to participate in either the market or the political process. A majority of those whose lives will be affected by the future climate are disfranchised because they do not yet exist. They possess no present-day economic power, and have no way of expressing their preferences in a tangible way. The only way to protect persons living in the future is through some kind of Intergenerational Constitution that spells out their rights, a Constitution that the present generation is committed to abiding by. Of course, it is always possible that we who are alive today will forsake our duty, and will make choices reflecting an attitude of domination over the world and the future rather than a sense of stewardship. The present generation always has the power to disregard the well-being of future generations or, for that matter, those who are powerless in the present day. This kind of power is an intrinsic element of human freedom, and the potential for its misuse is part of the tragedy of the human condition. Nevertheless, we have a responsibility to use our power wisely, in the hope that those who follow us will remember us with affection and honor rather than with an astonished dismay at our folly.

Notes

1 An Overview of the Issues

1. The IPCC was jointly established in 1988 by the World Meteorological Organization (WMO) and the United Nations Environment Programme (UNEP). It draws on the work of hundreds of experts from all regions of the world in presenting reports to the parties engaged in negotiations under the United Nations Framework Convention on Climate Change (UNFCCC), and represents the mainstream scientific consensus (or range of opinion where there is no consensus) on climate issues. The IPCC has produced to date three *Assessment Reports* (1990, 1995, and 2001) covering all aspects of climate change.
2. In the terminology of Working Group I of the IPCC, "very high confidence" means having a probability of 95 percent or more.
3. It should be noted that the statement in the preceding quotation that climate change will affect human welfare "positively and negatively" is based on the findings of some studies estimating that segments of the populations of certain countries might benefit from limited climate change.
4. Partial equilibrium analysis usually involves examination of only one market in isolation. The supply and demand curves of introductory economics courses are the icons of this approach.
5. Examples of climate policies analyzed include measures such as a tax on carbon emissions, a system of tradable emissions permits, or incentives to promote technical change to increase energy efficiency.
6. A fine historical overview is given by Ingrao and Israel (1990).
7. It is an open philosophical question why mathematical models work so well in the physical sciences. It can be argued that mathematics somehow corresponds to the nature of physical reality, or that logical thought evolved in conformity with physical reality because such thinking would have survival value. However, this view just pushes the question back a level – why should what has survival value to creatures operating on human-sized scales have applicability to the motions of planets and stars, or to the structure of invisible realms at the atomic level that can only indirectly be perceived by humans? It is clear that the current frontiers of physics – the paradoxes of quantum theory, the higher dimensions of string theory, and the outlines of alternative grand unification schemes – are so far from our ordinary experience that the concepts can only be formulated in abstract mathematical terms. For a classic discussion of the mystery of the effectiveness of mathematics in explaining physical phenomena, see Wigner (1960). As he puts it, "[t]he miracle of the appropriateness of the language of mathematics for the formulation of the laws of physics is a wonderful gift which we neither understand nor deserve. We should be grateful for it and hope that it will remain valid in future research and that it will extend, for better or for worse,

161

to our pleasure, even though perhaps also to our bafflement, to wide branches of learning." It has also been suggested that the success of mathematics in the natural sciences is partly the result of focusing only on those phenomena that can be described using traditional mathematical methods (see Wolfram 2002).

8. As Krugman says,

> academic economics . . . [is] a primitive science, of course. If you want a parallel, think of medicine at the turn of the [19th] century. Medical researchers had, by that time, accumulated a great deal of information about the human body and its workings, and were capable of giving some critically useful advice about how to avoid disease. They could not, however, cure very much. Indeed, the doctor/essayist Lewis Thomas tells us that the most important lesson from medical research up to that time had been to leave diseases alone – to stop the traditional "cures," like bleeding, that actually hurt the patients.   (1995, p. 9)

9. Strictly speaking, an "integrated assessment" model would have fully developed atmospheric, geophysical, and biological components coupled to socioeconomic components tracking population, emissions, and policy variables. In actuality, the general equilibrium models used for policy analysis are almost exclusively *economic* models, with the climate components included in only the most stylized way (through a "damage function" based on average surface temperature, for example). The term "integrated assessment" will be used to describe the models treated in this book, with the understanding that what is being discussed is the economic modeling and not the physical science side.

10. A "Pareto optimum" is an allocation of goods and services such that no individual's material situation (or welfare) can be improved without worsening the situation of at least one other person. Movements towards a Pareto optimum can, at least in principle, command social unanimity; movements away from a Pareto optimum cannot.

11. The modeling philosophy that will be carried throughout the following chapters is that the value of models is to illustrate arguments clearly and sharply. The models are not intended to be "descriptive" or all-inclusive. Indeed, one of the points that will be made again and again is that fully descriptive modeling is beyond the capacity of economics at the present time. There is also a difference between using models to claim to describe what *does* happen and using them to show what *might* happen. The former is the style of much current applied economics, including the general equilibrium climate policy models. The latter is what will be pursued in this book.

12. An interesting current example has to do with the nature of the Social Security entitlement. The Social Security Trust Fund has purchased trillions of dollars worth of US Treasury securities during the years it has been running a surplus. These purchases have enabled the federal government to have higher expenditures and lower taxes that would have been possible if the Social Security Administration had not been buying the Treasury bonds. At some point in the future, however, annual payroll tax receipts will no longer exceed Social Security payouts, and the Social Security system will have to

draw on first the interest, then the principal, of its Treasury bond holdings to meet the Social Security entitlements. The government could, of course, repudiate the bonds and declare that the entitlement was reduced. This might be politically very difficult, and it might fairly be described as immoral, but it is possible because of the sovereignty of the State.

2 The Representation of Consumers' Preferences and Market Demand

1. Competitive behavior also may diverge from economic rationality. A particularly striking example of this is the "Dollar Auction Game" first described by Martin Shubik (1971) and elaborated by Mérö (1998). In this game, a dollar is auctioned to the highest bidder, with the proviso that the second-highest bidder must also pay the auctioneer the amount that he bid. In both casual and formal experimental settings, the "winning" bid is typically considerably higher than $1.00. Shubik reported that in social settings, the average paid for the dollar was $3.40. Competitiveness, emotion, and a desire not to be made a fool of drive the bids considerably above the "economic" value of the dollar.

2. Ethical questions aside, Lane (2001) brings together research from a number of fields to make the case that once people are above the poverty level, additional income has little to do with happiness and that friendship and a good family life are the main sources of well-being in advanced economies. It is well-established that average reported levels of subjective happiness in a given country do not change very much as aggregate per capita income grows over time, although subjective happiness scores are positively related to income at a given point in time. See the survey by Frey and Stutzer (2002).

3. Indeed, this feature, coupled with the empirically testable notion that a market system free of discrimination will eventually lead to a distribution of wealth independent of the "initial conditions" (even an unjust "original" wealth distribution based on conquest), is the strongest argument for the justice of the market system. See DeCanio (1992) for a full discussion.

4. In particular, environmental externalities can be seen as arising from a lack of property rights, or from the difficulty of their enforcement due to high transactions costs. See Coase (1960), Dahlman (1979), Macaulay and Yandle (1977), and Dales (1968).

5. The idea that emissions rights should inhere primarily in the population as a whole has even been embodied in proposed legislation: the Clean Power Act that would reduce emissions of four major power plant pollutants including CO_2. The Clean Power Act as of June 2002 had 22 Senate and 128 House cosponsors (NRDC 2002).

6. To read an encompassing critique of CBA as it is currently practiced, see Heinzerling and Ackerman (2002).

7. Some economists have attempted to develop a theory of "rational addiction" to explain widespread destructive phenomena such as cigarette smoking and illegal drug use (e.g., Becker and Murphy 1988, Dockner and

Feichtinger 1993). Needless to say, this theory relies on a highly contestable concept of "rationality."

8. It is interesting that smokers on average have an unrealistically *high* estimate of their odds of dying from the effects of smoking (Viscusi 1991).

9. See Haddad and Howarth (forthcoming) and Diamond and Hausman (1994) for a full discussion of these and other conceptual and empirical flaws in the CV methodology. Needless to say, the validity of the CV methodology is hotly disputed. In addition to the Diamond and Hausman paper, the symposium on Contingent Valuation in the Autumn 1994 issue of the *Journal of Economic Perspectives* contains a spirited defense of CV by Hanemann (1994) and an overview by Portney (1994). In a recent contribution, Carson et al. (2001) conclude that "many of the alleged problems with CV can be resolved by careful study design and implementation" and that "claims that empirical CV findings are theoretically inconsistent are not generally supported by the literature" (p. 173). Consensus within economics on the validity and limits of CV has yet to be reached.

10. Obviously, the political rules matter. Some revenue-raising measures may require a supermajority (such as a two-thirds or three-fourths majority), and some property rights may be relatively inviolable on constitutional grounds.

11. Efforts to impose the rational maximization paradigm in nonmarket areas, such as the "economics of marriage and the family," the "economics of crime" (as well as the "economics of rational addiction") overreach and tend to confuse the fact that people respond to incentives with a full explanation of behavior. The two are quite different.

12. As Popper showed many years ago, statements that are tautologically true and therefore immune to empirical testing cannot constitute the foundation of any science; truly scientific propositions must at least in principle be falsifiable (Popper 1968 [1934]).

13. This situation may be changing. Economists have begun to draw on and contribute to the extensive literature that deals scientifically with human actions that do not fit the strict definition of economic rationality. See Laitner et al. (2000) and the references cited therein, or the session of the 2000 Annual Meeting of the American Economic Association titled "Preferences, Behavior, and Welfare" that had papers devoted to "Emotions in Economic Theory and Economic Behavior" (Loewenstein 2000) and also "Thinking and Feeling" (Romer 2000). The National Bureau of Economic Research currently has a Working Group on Behavioral Finance (NBER 2002). The 2002 Nobel prize in economics was awarded in part for work in behavioral economics. Whether or not this approach will find its way into integrated assessment modeling remains to be seen.

14. There have been some interesting efforts to derive the weakest set of conditions on individual utility functions or the distribution of individual characteristics that would imply well-behaved aggregate demand functions. See, for example, Hildenbrand (1994), Härdle et al. (1991), Freixas and Mas-Colell (1987), and Jerison (1982). This important work has had essentially zero impact on the practice of integrated assessment climate modeling. As will be shown below, the empirical studies on the way environmental goods enter individuals' preferences that would need to be done to bring to bear on climate models the insights of the theorists remain undone, and

may be impossible to carry out in the absence of property rights and markets in the key environmental goods.

15. There is a correspondence between the axiomatic and utility function approaches, with major points of overlap and some minor technical differences. A detailed treatment is given by Mas-Colell et al. (1995).

16. See Ackerman (1998), Laitner et al. (2000), or Mas-Colell et al. (1995). Ingrao and Israel (1990) place the development of these ideas in the context of the history of economic (and scientific) thought on the subject of equilibrium. The result that aggregate market demand curves can have virtually any shape is usually referred to as the Sonnenschein–Mantel–Debreu theorem after the theorists who wrote the pioneering papers.

17. Here is an indicative example: the handbook *Applying General Equilibrium* (Shoven and Whalley 1998 [1992]), in the respected Cambridge Surveys of Economic Literature series, makes no mention of "uniqueness" or "stability" in its index. The authors do allude to the fact that some numerical methods for solving the systems of equations have convergence problems (pp. 41–2), and they acknowledge that "[t]here are no guarantees that such procedures [price adjustment processes in which prices are raised for commodities with positive excess demands and lowered for commodities with excess supplies] will converge because the excess demand functions may have local inflection points, changes in slope, or both . . . whence our need for a more elaborate computational method of finding counterfactual equilibria" (p. 39). Whether or not "a more elaborate computational method of finding . . . equilibria" has any correspondence to economic reality is not examined. This point will be developed in the following section.

18. Analogous results can be obtained in models with production. Indeed, the possibilities for multiple equilibria in economies with production may be greater than in exchange economies. Timothy Kehoe (1985) says,

> Unfortunately, the conditions required for uniqueness of equilibrium in production economies appear to be more restrictive than those in pure exchange economies. For example, it is well-known that, if either the weak axiom of revealed preference or gross substitutability is satisfied by the consumer [aggregate] excess demand function, then a pure exchange economy has a unique equilibrium. This is not the case for an economy with production. (pp. 120–1)

Aside from their simplicity, there is an additional reason to restrict our attention to exchange models rather than models involving production. Intertemporal models with production typically focus on the choice between saving and consumption. However, this makes the choices and opportunities of future generations contingent on the savings/consumption decisions of the present generation, because the capital available to the future generations will depend on how much saving is undertaken by the present generation. If the analysis means to treat the generations in an ethically neutral fashion (with no generation having an analytically preferred vantage point), the endowments of each generation should be in some sense independent of the choices and preferences of the others. This condition is most easily represented in exchange models. Intertemporal models of this type will be developed more fully in Chapter 3. It should be noted that generalizations of these ethically neutral exchange models are possible if the endowments change over time exogenously, as is the case if "autonomous" technical progress is providing the impetus for economic growth.

19. If the benefit/cost ratio differs from one at the imputed price of the environmental goods, then the economy cannot be in equilibrium; the environmental policy has to be implemented until the ratios of marginal utility to price are equalized across all goods (including the environmental goods).

20. The setup of this model economy will follow the example given by Kehoe (1998). Kehoe illustrated multiple equilibria in an exchange economy with only two agents and two goods; the examples developed subsequently in the text are straightforward generalizations.

21. Obviously, there is nothing intrinsic about this assumption of an equal number of agents and goods. It is adopted purely for expositional convenience.

22. It should be noted that in the real economy, not only do different pairs (or combinations) of goods have different degrees of substitutability, but those elasticities of substitution may change over time. The degree of substitutability between goods may also differ with the consumer's income level.

23. In the limit as $b_i \to 0$, the utility function of equation (2.1) becomes $u_i = \Sigma_j a_{ij} \log(x_{ij})$. This can be seen by subtracting one from the numerator of each of the summed terms in equation (2.1) (a monotonic transformation of a utility function yields an equivalent utility function) and applying L'Hospital's rule to calculate the limit. In the Cobb–Douglas utility function ($u_i = \Pi_j x_{ij}^{a_{ij}}$), the elasticity of substitution between goods is one; in the Leontief utility function ($u_i = \min[(x_{i1}/\beta_{i1}), (x_{i2}/\beta_{i2}), \ldots , (x_{in}/\beta_{in})]$), the elasticities of substitution are zero. Note that $u_i = \Sigma_j \alpha_{ij} \log(x_{ij})$ is the same as Cobb–Douglas utility, because the two utility functions differ only by a monotonic transformation (in this case, taking logarithms of the Cobb–Douglas function).

24. Substitutes and complements can also be defined in terms of the response of the demand for one good to a change in the price of the other. That is, if the demand function for (say) the first good is given by $h(p_1, p_2)$, then the two goods are complements (at a particular pair of values of the prices) if $\partial h(p_1, p_2)/\partial p_2 < 0$; they are substitutes if $\partial h(p_1, p_2)/\partial p_2 > 0$. The partial derivative $\partial h(p_1, p_2)/\partial p_2$ can be converted into an elasticity by multiplying by $p_2/h(p_1, p_2)$, and the resulting "cross-elasticity of demand" will have the same sign as the partial derivative. To simplify both the mathematics and the exposition, the models used in the text will be of the constant elasticity of substitution type, so that the degree of substitutability of the commodities depends exclusively on the b_i parameters as in equation (2.3).

25. This normalization rule is adopted to facilitate the numerical search for equilibria. With the prices restricted to lie between zero and one, the starting points for the numerical search for equilibria can be confined to a well-defined region. The substantive results are the same if a numeraire good is selected and its price is set equal to one, because the relative prices are the same in either case.

26. The general equilibrium system was solved numerically in Mathematica (Wolfram 1999).

27. For this exchange economy, GDP is $\Sigma_j p_j(\Sigma_i x_{ij}) = \Sigma_j p_j(\Sigma_i \omega_{ij})$. The constancy of GDP follows from the symmetry of the endowments (so that $\Sigma_i \omega_{ij}$ is the same for all j) and the fact that the prices are normalized to sum to one.

28. Note that with five agents, the Gini for the most extreme degree of inequality (all the income accruing to a single agent) is 0.8.

29. The fact that the equilibria of a competitive market economy are Pareto optima is just the first fundamental welfare theorem. As noted earlier, it requires an absence of externalities, perfect information, no collusion, and the other standard assumptions of the competitive general equilibrium model.

30. This is essentially the same as the instability of prototypical zero-sum game of "Couples" discussed by Riker and Ordeshook (1973). In real politics, other preferences not reflected in the model may determine which majority coalition actually is formed.

31. Kehoe (1998) shows how the index theorem can be used to derive necessary and sufficient conditions for the uniqueness. See also note 39.

32. In this case Newton's method exhibited convergence problems from some starting points. It is possible that there are more equilibria in this economy, but for many different initial price vectors Newton's method did not converge, even after thousands of iterations.

33. For purely mathematical reasons, the number of equilibria must always be odd. See Dierker (1972).

34. In particular, Mathematica's FindRoot procedure was used with different starting points. In this application (in which formulas for the derivatives of the equations being solved simultaneously can be calculated explicitly), FindRoot uses Newton's method or the secant method (Wolfram 1999, pp. 919–22).

35. Exceptions, such as the Goulder–Schneider model (1999), rely on ad hoc adjustment processes that are akin to the tâtonnement that will be described presently.

36. Considerable effort from the time of Walras and Marshall onward (see Ingrao and Israel 1990) has been expended to develop dynamic theories that are consistent with the equilibrium model. Despite all the good work that has been done, Fisher's assessment (1989) indicates how much remains: "[W]e have no rigorous basis for believing that equilibria can be achieved or maintained if disturbed.... [T]here is no disguising the fact that this is a major lacuna in economic analysis" (p. 36), and "[N]ew modes of analysis are needed if equilibrium economic theory is to have a satisfactory foundation" (p. 42).

37. According to Hahn (1989), the Walrasian "auctioneer" calls out hypothetical prices in one market at a time and adjusts them according to the excess demand. When equilibrium is reached in one market, the auctioneer moves on to the next market. Interestingly, Fisher, in the same *New Palgrave* volume (1989), states that "[t]he question of who adjusts prices . . . is typically left unanswered or put aside with a reference to a fictitious Walrasian 'auctioneer'. That character does not appear in Walras (who did have prices adjusting to excess demands) but may have been invented by Schumpeter in lectures and introduced into the literature by Samuelson . . ." (p. 37).

38. Herings (1998) offers a more optimistic assessment of the success of the effort to relate computational algorithms with nice convergence properties to actual economic adjustment processes.

39. Kehoe (1998) describes a regular economy as follows:

> First, a regular economy has a finite number of equilibria. . . . Second, each equilibrium of a regular economy varies continuously with the underlying characteristics of the economy. Third, in the set of all possible economies given an appropriate topological structure, almost all economies are regular. Fourth, we can use a fixed point index theorem to develop necessary and sufficient conditions for uniqueness of equilibria of regular economies. (pp. 47–8)

40. See Ebert (2001) for a discussion of the kinds of severe and restrictive assumptions that have to be made to evaluate preferences for a nonmarket good even if the demand system for the other goods is known. As Ebert says,

> it is generally impossible to recover the underlying (unconditional) preference ordering over commodity bundles containing the private goods and the nonmarket good. In particular, neither the total nor the marginal willingness to pay for the nonmarket good can be inferred from these data since there is always an infinite variety of unconditional preference orderings implying the behavior observed. (p. 374)

3 The Treatment of Time

1. "For a full coupled atmosphere/ocean GCM [General Circulation Model], . . . the heat exchange with the deep ocean delays equilibration and several millennia, rather than several decades, are required to attain it" (IPCC 2001a, p. 533).

2. The calculation is based on a present-day world GDP of $30 trillion and population of 6 billion. The present value of the world income stream from a point τ in the future is given by

$$\text{NPV} = \sum_{t=\tau}^{\infty} \frac{\text{GDP}_0(1+g)^t}{(1+r)^t} = \text{GDP}_0\left(\frac{1+g}{1+r}\right)^\tau\left(\frac{1+r}{r-g}\right)$$

where GDP_0 is world GDP today, g is the assumed growth rate of world GDP, and r is the assumed market rate of interest, provided $r > g$. But can there be circumstances under which r is less than or equal to g? In such cases, the NPV does not converge.

3. For further discussion, see Arrow (1999) and Koopmans (1960, 1965).

4. According to the discounted utility formulation, an individual's subjective time rate of discount δ is determined by the relationship $x = (1 + \delta)^t$, where the individual would be indifferent between $1 in consumption today and $x of consumption at time t in the future. (The continuous-time version of the same formula is $x = e^{\delta t}$.)

5. Among the anomalies demonstrating the inadequacies of the DU model as a descriptive model of behavior are hyperbolic discounting (the tendency of implicit discount rates over long horizons to be lower than the implicit discount rate over short horizons), the "sign effect" (gains are discounted more than losses), the "magnitude effect" (small outcomes are discounted more than large ones), the "delay-speedup" asymmetry (respondents demand more to expedite payment than they would pay to delay it), the preference for improving sequences, and violations of independence and preference for spread (Frederick et al. 2002).
6. The non-participation of people living in the future is the main reason Broome (1992) rejects "discounting at the consumer interest rate."
7. A concise summary of Arrow–Debreu general equilibrium theory is Geanakoplos (1989a).
8. The phrase was introduced by Graciela Chichilnisky. See her article characterizing sustainable development axiomatically (1997).
9. This elasticity of the marginal utility of consumption is defined as $\alpha = -(c/u'(c))(du'(c)/dc)$, where u is the utility function and c is consumption in any given time period so that $u = (a/b)c^b$.
10. Alternatively, the derivations could be carried out in continuous time.
11. The quotation is specifically in reference to the point that equilibrium values computed depend as much on numerical technology (solution algorithm) as on the starting point.
12. In the real world, of course, these substitution parameters would all be different – across goods, time periods, and individuals. The assumption that they are all the same is retained here because the purpose of the model is illustrative, rather than attempting to replicate reality. But note the empirical burden of any modeler who would purport to present a "realistic" model that captures all the key features of the economy.
13. Appropriate scaling can reduce the number of parameters in this case to 13.
14. A richer model could be constructed in which the endowments change over time.
15. This value is from Auerbach and Kotlikoff (1987).
16. Think of individuals beginning their economic lives at 21 and living into their eighties.
17. Kehoe and Levine (1990) derived analogous results.
18. The (arbitrarily chosen) parameters for this $2 \times 2 \times 2$ Cobb–Douglas example are: $a_{11} = 3$, $a_{12} = 1$, $a_{13} = 1$, $a_{14} = 1$, $a_{21} = 2$, $a_{22} = 1$, $a_{23} = 2$, $a_{24} = 1$, $\omega_{10} = 60$, $\omega_{11} = 70$, $\omega_{20} = 5$, $\omega_{21} = 10$, $\varepsilon_{10} = 60$, $\varepsilon_{11} = 5$, $\varepsilon_{20} = 5$, $\varepsilon_{21} = 5$, and $\lambda = 0.5$.
19. The parameters of this utility function are: $b = -4$, $a_{11} = 512$, $a_{12} = 1$, $a_{13} = 512$, $a_{14} = 1$, $a_{21} = 1$, $a_{22} = 512$, $a_{23} = 1$, $a_{24} = 512$, $\omega_{10} = 60$, $\omega_{11} = 5$, $\omega_{20} = 5$, $\omega_{21} = 60$, $\varepsilon_{10} = 60$, $\varepsilon_{11} = 5$, $\varepsilon_{20} = 5$, $\varepsilon_{21} = 60$. The speed of adjustment $\lambda = 0.5$. Because of the particular parameters of this utility function, the equilibrium ratio of the price of the environmental good to the ordinary good is always 1. The numerical search for equilibrium in each period has two potential starting points, $\beta = 0.1$ or $\beta = 1.1$. (The starting value of ϕ is 0.9 in either case.) The probability of picking the first of the two starting points is 0.7; the probability of picking the second starting point is 0.3. These parameters are not meant to be "realistic"; they are illustrative of what can happen in the OLG model with multiple steady states.

20. Newton's method as implemented by the FindRoot command in Mathematica.

21. There are other ways of avoiding an infinite sum. Growth models can be set up in which savings are determined by some fixed rule, such as being a constant fraction of output. In that case, a savings rate can be determined that maximizes steady-state consumption at each point in time. This steady-state path is referred to as the "Golden Rule" growth path because it corresponds to the path with the highest *sustainable* per capita consumption. Other paths could exhibit higher per capita consumption for early generations only at the expense of the standard of living of the later generations, hence the moral connotations of the "Golden Rule" path. See Solow (1970).

22. Mathematically, let $u_i(x_{i1}, x_{i2}, \ldots, x_{in})$ be individual i's utility function, where x_{ij} is the consumption of good j by individual i, and suppose $g[\cdot]$ is any function such that $g > 0$ and $g' > 0$. The individual's demand functions will be identical whether u or $g[u]$ is used as the utility function; that is, all the empirical implications of utility maximization are the same whether u or $g[u]$ is maximized. The basic reason is that the first-order conditions for utility maximization are that the ratios of marginal utilities be equal to the corresponding price ratios. By the chain rule,

$$\frac{\partial g[u_i(x_{i1}, x_{i2}, \ldots, x_{in})]/\partial x_{ij}}{\partial g[u_i(x_{i1}, x_{i2}, \ldots, x_{in})]/\partial x_{ik}} = \frac{g'[u_i](\partial u_i/\partial x_{ij})}{g'[u_i](\partial u_i/\partial x_{ik})} = \frac{\partial u_i/\partial x_{ij}}{\partial u_i/\partial x_{ik}} = \frac{p_j}{p_k}$$

23. See Cooter and Rappoport (1984), who argue that the "the ordinalist revolution [of the 1930s] represented a change, not progress in economics" (p. 508).

24. The result of maximization of (3.35) is invariant under positive linear transformations of the utility function, but not under arbitrary positive monotonic transformations. See Barro and Sala-i-Martin (1995) and Koopmans (1965).

25. It needs to be kept in mind that various transformations of the same underlying parameter have different names. As noted earlier, the elasticity of the marginal utility of consumption is the inverse of the intertemporal elasticity of substitution for CES utility functions of the type being considered here.

26. The coefficient of relative risk aversion (usually designated by α) in the CES utility function is equal to $1 - b$, so it is the inverse of the intertemporal elasticity of substitution for these utility functions.

27. In Kocherlakota's words, "for any such [value of the coefficient of relative risk aversion less than 8.5], the representative agent can gain at the margin by borrowing at the Treasury bill rate and investing in stocks" (1996, p. 49, footnote omitted). Kocherlakota goes on to argue that while "some economists" believe there is no equity premium puzzle (because "individuals are more risk averse than we thought"), a "vast majority of economists believe that values for α above ten (or, for that matter, above five) imply highly implausible behavior on the part of individuals." At the same time, Kocherlakota cites Kandel and Stambaugh (1991) who show that "even values of α as high as 30 imply quite reasonable behavior when the bet involves a maximal potential loss of around one percent of the gambler's

wealth" (1996, p. 52). Needless to say, the debate on the equity premium puzzle is still very much open. It is a leading example of a major empirical anomaly that the economics profession has not been able to resolve despite nearly two decades of intensive controversy and investigation.

28. Barsky et al. (1997) also found no relationship between the elasticity of intertemporal substitution and risk tolerance (the inverse of the coefficient of relative risk aversion) in their sample. Of course, the relationship between the two parameters depends on the form of the utility function, which is unknown.

29. The crucial importance of what kind of assumptions are made about the substitutability or complementarity of environmental goods (or natural capital) and marketed goods (or produced capital) has been demonstrated by Neumayer (1999). He argues that substitutability is more fundamental than the discounting for integrated assessment, and that "there is hardly any reliable empirical evidence" (p. 40) on the substitutability between ordinary consumption and environmental goods in people's utility functions.

4 The Representation of Production

1. As pointed out in Chapter 2, this does not mean that the multiple equilibria cannot exist in models with production.

2. For statement of the Porter hypothesis, see Porter (1991) and Porter and van der Linde (1995a, b). For a clear presentation of the conventional objections, see Palmer et al. (1995). For a recent model showing how Porter-type results may emerge in a richer model of production, see DeCanio et al. (2001a).

3. Coase (1937) raised the question of why we observe firms rather than markets mediating all transactions. This basic question has led to an extensive and productive intellectual exploration. See Alchian and Woodward (1988) for a review.

4. In some of the models, a linear programming framework is employed instead of the production function representation. The same arguments being made in the text apply.

5. Even in this extreme case, it is possible that individual members of the firm gain more from "resume-building" activities that are not particularly helpful to the organization than they would from devoting themselves wholeheartedly to the organization's well-being. The *Wall Street Journal* reported that even some employees laid off by the Enron bankruptcy felt good about their Enron experience. As one former employee put it, "I was well trained at Enron and I'll land on my feet" (Barrionuevo 2001).

6. This author's own contributions include DeCanio (1993, 1994a, b, c, 1998, 2000b), which also contain numerous references to the wider literature.

7. For a survey of antecedents, see Williamson (1964). A biblical example is the Parable of the Wily Manager:

> A rich man had a manager who was reported to him for dissipating his property. He summoned him and said, "What is this I hear about you? Give me an account of your service, for it is about to come to an end." The manager thought to himself, "What shall I do next? My employer is

sure to dismiss me. I cannot dig ditches. I am ashamed to go begging. I have it! Here is a way to make sure that people will take me into their homes when I am let go."

So he called in each of his master's debtors, and said to the first, "How much do you owe my master?" The man replied, "A hundred jars of oil." The manager said, "Take your invoice, sit down quickly, and make it fifty." Then he said to a second, "How much do you owe?" The answer came, "A hundred measures of wheat," and the manager said, "Take your invoice and make it eighty."

The owner then gave his devious employee credit for being enterprising! Why? Because the worldly take more initiative than the other-worldly when it comes to dealing with their own kind.

(The New American Bible 1970, Luke 16: 1–8)

8. Appeal to an evolutionary argument that the firm will act "as if" it were maximizing profits is insufficient; the "as if" argument first proposed by Friedman (1953) has been shown to rest on a number of strong and non-established subsidiary hypotheses (Laitner et al. 2000). See below (section 4.4) for a discussion of what a real evolutionary dynamic might yield.

9. This will hold for any kind of "classical" computer (i.e., one equivalent to a Turing machine). It is one of the outstanding problems of mathematics to show whether no such algorithm exists – or that one just has not been discovered yet. This "**P** vs **NP**" problem is one of the seven problems for which the Clay Mathematics Institute has offered a $1 million prize for a valid solution. (Details and conditions are given at www.claymath.org/prizeproblems/index.htm.) Also, while quantum computers can in principle solve some problems that are intractable for classical computers (such as factorization of large integers), it is not known whether a quantum computer could solve problems in the **NP**-complete class in polynomial time (Shor 1997).

10. Formal computational complexity theory is not required to arrive at the ubiquity of bounded rationality in economics and other realms of human activity, of course. Herbert Simon did the pioneering work for which he won the Nobel prize without reference to it. See Conlisk (1996) for a comprehensive review of Simon's and subsequent work on bounded rationality.

11. Leibenstein (1966). These papers through 1989 have been collected in Button (1989), but Leibenstein continued to explore the issues through the 1990s (Leibenstein and Maital 1992, 1994).

12. This bibliography excludes working papers and technical reports.

13. Førsund (1999) contains a discussion of the methodologies.

14. See Karpoff and Malatesta (1989) and the literature they cite; also Schranz (1993). There are, however, economists who doubt that antitakeover legislation has any effect on shareholder wealth (Pugh and Jahera 1990). Harris (1990) argues from agency theory that the shareholders of takeover targets may be willing to allow their managers to adopt antitakeover measures. A "golden parachute" may enable the managers of the target firm to bargain harder for a share of the synergistic gains from the potential merger, because they have some protection against the loss of their jobs that will result if the takeover attempt is successful.

15. The Fisher–McGowan result was the subject of considerable controversy when it appeared (Long and Ravenscraft 1984, Horowitz 1984, Martin 1984, Van Breda 1984). Fisher effectively demolishes these criticisms in his rebuttal (Fisher 1984), and notes that other authors, most notably Harcourt (1965), had anticipated Fisher and McGowan.

16. Efficiency is defined here as the condition that production is organized to get the maximum output achievable from the given factors of production. Fisher observes that "mere identity of technologies and constant returns does *not* imply the existence of an aggregate production function as may be seen by trying to add up two identical Cobb–Douglas production functions without further restrictions" (1969b, p. 556).

17. For example, the *only* condition under constant returns in which a capital aggregate exists is that of purely capital augmenting technical differences. See Fisher (1969b) and the references he cites.

18. Nelson (1995) provides a review; see also Dosi (2000). Hodgson (1993) provides a historical survey and analysis. The *Journal of Evolutionary Economics* has been in operation since 1990.

19. In addition to the references cited in these papers, see DeCanio and Watkins (1998) and DeCanio et al. (2001b).

20. In general, the vertices can also represent subunits or divisions of a larger organization.

21. The number of labeled digraphs of size n is $2^{n(n-1)}$. Two undirected graphs G_1 and G_2 are isomorphic if there is a one-to-one correspondence between the vertices of G_1 and those of G_2, with the property that the number of edges joining any two vertices of G_1 is equal to the number of edges joining the corresponding vertices of G_2. Two digraphs are isomorphic if there is an isomorphism between their underlying (undirected) graphs which preserves the ordering of the vertices in each edge (Wilson 1985).

22. Two-point crossover is defined as follows. Pick two points at random in the chromosome string. Then if the two parents' chromosomes are the strings ABC and DEF (with the two randomly selected points being the dividing points between strings A and B and strings B and C, respectively, for the first parent, and between strings D and E and strings E and F for the second parent), then the two offspring of these parents will have chromosomes AEC and DBF.

23. If all the members of the population have a positive fitness score, an organization's relative fitness is simply the organization's fitness score divided by the sum of the scores for all the members of the population. If the least fit member of the population has a negative fitness score, it is necessary to scale the scores by subtracting this negative value so that all the scaled scores are nonnegative. With subpopulations of different species, there is an added wrinkle if the species subpopulation's least fit member has a negative score – the scaling used to select the second parent may be based either on the entire population's least fit member or the species' least fit member. Unless stated otherwise, the convention followed in the results reported here will be based on scaling using the population's least fit member for scaling.

24. Histograms and sample statistics were computed using EViews (Quantitative Micro Software 1997).

25. It is also clear in Figure 4.3 and other instances not shown here that the distribution of fitness scores is not normal. Jarque-Bera tests indicate rejection of the null hypothesis of normality at very high confidence levels. The same conclusion follows if the adjusted Lagrange multiplier (ALM) test for normality (Urzúa 1996) is conducted.

26. The test statistics for the population elite runs are larger than for the others because population elite selection almost always resulted in final populations with no variation in their fitness scores. For example, 94 of the 100 size-15 runs exhibited no variation in their final populations. These outcomes obviously cannot be coming from the same underlying distribution.

27. Note that the type of path dependence influencing the evolutionary dynamic stems from the fact that each generation can only draw on the genetic material surviving in the previous generation. This is distinct from path dependence arising because increasing returns to scale give some producers a self-reinforcing advantage over others (Arthur 1994).

28. Of course, organizational demography is a burgeoning field that extends far beyond the boundaries of economics. See Carroll and Hannan (2000) for a recent synthesis.

29. See Canan and Reichman (2001) and Reichman et al. (2001) for an analysis of the importance of leadership and intragroup network structures in the success of the Montreal Protocol on Substances that Deplete the Ozone Layer.

5 The Forecasting Performance of Energy-Economic Models

1. It could be argued that an incautious willingness on the part of environmentalists to warn that "the sky is falling" in the face of all types of pollution, regardless of how serious, has reduced the credibility of the entire environmental movement.

2. The organizations polled for *Blue Chip Economic Indicators* include the UCLA Business Forecasting Project, DRI-WEFA, the Conference Board, and the US Chamber of Commerce (Moore 2002).

3. Lovins, personal communication (1998).

4. Nash sought to use the original testimony or submission of participants in the debate for reprint in his collection. NERA agreed to inclusion of the Perry and Streiter paper only on condition that a revised version of the paper (prepared after the congressional testimony) be printed, along with a memo responding to Lovins' comments on their critique.

5. This author was a participant in the interagency process while serving as a Senior Staff Economist at the Council of Economic Advisers in 1986–87, and can testify to the considerable effort and resources that were devoted to this assessment.

6. This DOE study was done as part of the National Energy Policy Plan (NEPP) of 1983.

7. The forecasts were for 1990 projections contained in the AEOs of 1983, 1985, and 1987. There were 182, 185, and 177 energy sectors in these three AEOs, respectively.

8. These large forecast errors were found even after about 50 sectoral forecasts were discarded from the sample because their x values exceeded 100. As Shlyakhter et al. observe, "[w]e assumed that the AEO model might not be applicable in those cases and omitted them . . ." (1994, p. 123).

9. Storage of these fossil fuels can easily take the form of leaving known reserves in the ground rather than extracting them.

10. The spot price also requires specification of grade and location of the commodity.

11. A 1995 study by the Office of Technology Assessment (OTA) also found a systematic tendency to overestimate *ex ante* the cost of OSHA regulations (US Congress 1995).

12. See Harrington et al. (2000). They cite the testimony of EPA Administrator Carol Browner before the US Senate Committee on Environment and Public Works (1997). See also Bohi and Burtraw (1997).

13. The coefficient of variation of a sample is the ratio of the sample standard deviation to the sample mean. It is a units-free measure of variability in a sample.

14. The studies included one by the US EIA (1998b), one by the Council of Economic Advisers (1998b), a synthesis of the models participating in Energy Modeling Forum-16 (EMF 1999), and two by the Interlaboratory Working Group of five of the US national laboratories (1997, 2000).

6 Principles for the Future

1. Russian "hot air" is the gap between current emissions and the 1990 benchmark of the Kyoto Protocol. Current Russian emissions are considerably below their 1990 levels because of the restructuring of the Russian economy following the collapse of the Soviet regime.

2. These concerns are partially reflected in Metz et al. (2000).

3. It is a cliché that voters in the rich countries do not care much about the fate of people in the poor countries, but a richer and environmentally secure world, with the enhanced trading opportunities and political stability that would be part of it, is fundamentally in the interest of the citizens of the economically developed countries.

4. It should always be kept in mind that ultimately avoiding dangerous anthropogenic interference with the climate system "could require efforts, perhaps international, pursued with the urgency of the Manhattan Project or the Apollo space programme" (Hoffert et al. 1998, p. 884). Stabilization of the atmosphere at the 450 ppm CO_2 that may be needed to prevent thermohaline circulation shutdown and sea level rise from disintegration of the West Antarctic ice sheet (O'Neill and Oppenheimer 2002) could require 25 TW of emission-free power by 2050. Primary power consumption today is ~12 TW, of which 85 percent is fossil fueled (Hoffert et al. 2002).

5. For example, international debt relief for the poorest countries whose debts were incurred by irresponsible or undemocratic governments is a proposal that deserves the most serious consideration.

6. The flexibility association with the creation of new rights could also enable some of the harsher burdens of the transition away from fossil fuels in the

developed countries to be ameliorated – a share of the carbon rights (or revenues from a carbon tax or equivalent permit auction) could cushion the displacement of coal miners or other fossil fuel-dependent workers (Krause et al. 2003).

7. Appeals to the Golden Rule of economic growth theory have been made in policy debates in the past. In the Solow–Swan growth model with savings a fixed proportion of output (see Chapter 3, note 19), the "Golden Rule" savings rate that maximizes steady-state per capita consumption is one in which the interest rate just equals the rate of growth of per capita income (Solow 1970). If "willingness to pay" to avoid environmental damage has an income elasticity of unity (so that it grows at the same rate as per capita income), the interest rate appropriate for calculating the costs and benefits of investments to protect the environment is zero. This line of reasoning was proposed during the internal US government debate on ozone layer protection policy prior to the signing of the Montreal Protocol, and although it was not adopted officially by the government, low interest rates were used to approximate intergenerational neutrality in calculating the costs and benefits of ozone layer protection. These calculations showed that the benefits of ozone layer protection were orders of magnitude greater than the costs (Crawford 1987). For a full discussion, see DeCanio (2002).

8. It has been noted, however, that economics graduate students tend to behave more like the purely self-interested denizens of economic models than other people (Marwell and Ames 1981).

9. Here is the full text of the Economists' Statement:

> 1. The review conducted by a distinguished international panel of scientists under the auspices of the Intergovernmental Panel on Climate Change has determined that "the balance of evidence suggests a discernible human influence on global climate." As economists, we believe that global climate change carries with it significant environmental, economic, social, and geopolitical risks, and that preventive steps are justified.
> 2. Economic studies have found that there are many potential policies to reduce greenhouse-gas emissions for which the total benefits outweigh the total costs. For the United States in particular, sound economic analysis shows that there are policy options that would slow climate change without harming American living standards, and these measures may in fact improve U.S. productivity in the longer run.
> 3. The most efficient approach to slowing climate change is through market-based policies. In order for the world to achieve its climatic objectives at minimum cost, a cooperative approach among nations is required – such as an international emissions trading agreement. The United States and other nations can most efficiently implement their climate policies through market mechanisms, such as carbon taxes or the auction of emissions permits. The revenues generated from such policies can effectively be used to reduce the deficit or to lower existing taxes.

This statement was signed by over 2500 members of the American Economic Association, including eight US Nobel laureates in economics. The cover letter soliciting signatures to the statement was signed by Kenneth Arrow of Stanford, Robert Solow of MIT, Dale Jorgenson of Harvard, William Nordhaus of Yale, and Paul Krugman (then of MIT, now at Princeton) (DeCanio 1997).

10. Examples of the latter abound. Part of the reason for the proliferation of gas-guzzling passenger vehicles in the US is that "light trucks" were excluded from the original Corporate Average Fuel Economy (CAFE) standards. The light truck exemption was meant to avoid unnecessarily burdening small businesses (painters, gardeners and the like) that use pickup trucks. The exemption turned out to be a giant loophole through which the automobile makers drove their sport utility vehicles (SUVs). As a result, despite significant technological progress in engines and design, the fuel efficiency of the passenger vehicle fleet in the United States has been declining in recent years.

11. This approach is sometimes referred to as "backcasting." For example, see Robinson (1988).

12. This is the Kaya identity, that CO_2 Emissions = (CO_2 Emissions per unit of energy) $\times$ (Energy per unit output) $\times$ (Output per capita) $\times$ Population. It is found in Kaya (1989) and is reproduced in IPCC (1996).

References

ABARE (Australian Bureau of Agricultural and Resource Economics, 2002). *Global Trade and Environment Model (GTEM): a Computable General Equilibrium Model of the Global Economy and Environment*. Canberra: http://www.abare.gov.au/resear.ch/GTEM/gtem.doc.

Ackerman, Frank, 1998. "Still Dead after All These Years: Interpreting the Failure of General Equilibrium Theory," unpublished manuscript, Global Development and Environment Institute, Tufts University.

Alchian, Armen A., 1950. "Uncertainty, Evolution, and Economic Theory," *Journal of Political Economy*, Vol. 58, No. 3 (June): 211–21.

Alchian, Armen A., and Susan Woodward, 1988. "The Firm is Dead; Long Live the Firm: a Review of Oliver E. Williamson's *The Economic Institutions of Capitalism*," *Journal of Economic Literature*, Vol. 26, No. 1: 65–79.

Amano, Akihiro, 1997. "On Some Integrated Assessment Modeling Debates," paper presented at the IPCC Asia-Pacific Workshop on Integrated Assessment Models, United Nations University, Tokyo, March 10–12.

Amir-Atefi, Keyvan, 2001. "Bounded Rationality, Rules of Thumb, and Organizational Design," unpublished manuscript (October 16).

Antle, Rick, and Gary D. Eppen, 1985. "Capital Rationing and Organizational Slack in Capital Budgeting," *Management Science*, Vol. 31, No. 2 (February): 163–74.

Arrow, Kenneth J., 1999. "Discounting, Morality, and Gaming," in *Discounting and Intergenerational Equity*, eds Paul R. Portney and John P. Weyant. Washington, DC: Resources for the Future.

Arrow, Kenneth J., H.D. Block, and Leonid Hurwicz, 1959. "On the Stability of the Competitive Equilibrium, II," *Econometrica*, Vol. 27, No. 1 (January): 82–109.

Arthur, W. Brian, 1994. *Increasing Returns and Path Dependence in the Economy*. Ann Arbor: The University of Michigan Press.

Ascher, William, 1978. *Forecasting: an Appraisal for Policy-Makers and Planners*. Baltimore: The Johns Hopkins University Press.

Athanasiou, Tom, and Paul Baer, 2001. "Hello World," *Climate Equity Observer*, No. 1, http://www.ecoequity.org

Auerbach, A.J., and L.J. Kotlikoff, 1987. *Dynamic Fiscal Analysis*. Cambridge: Cambridge University Press.

Australian Bureau of Agricultural and Resource Economics (ABARE), 1996. The MEGABARE model: interim documentation. Canberra: Australian Bureau of Agricultural and Resource Economics.

Baer, P., J. Harte, B. Haya, A.V. Herzog, J. Holdren, N.E. Hultman, D.M. Kammen, R.B. Norgaard, and L. Raymond, 2000. "Equity and Greenhouse Gas Responsibility in Climate Policy," *Science*, Vol. 289: 2287.

Barnes, Peter, 2001. *Who Owns the Sky?: Our Common Assets and the Future of Capitalism*. Washington, DC: Island Press.

Barnes, Peter, and Rafe Pomerance, 2000. *Pie in the Sky: the Battle for Atmospheric Scarcity Rent*. Washington: Corporation for Enterprise Development.

Barrionuevo, Alexi, 2001. "Jobless in a Flash, Enron's Ex-Employees Are Stunned, Bitter, Ashamed," *Wall Street Journal*, 12/11/01: B1, B12.

Barro, Robert J., and Xavier Sala-i-Martin, 1995. *Economic Growth*. New York: McGraw-Hill, Inc.

Barsky, Robert B., F. Thomas Juster, Miles S. Kimball, and Matthew D. Shapiro, 1997. "Preference Parameters and Behavioral Heterogeneity: an Experimental Approach in the Health and Retirement Study," *The Quarterly Journal of Economics*, Vol. 112, No. 2 (May): 537–79.

Becker, Gary S., and Kevin M. Murphy, 1988. "A Theory of Rational Addiction," *The Journal of Political Economy*, Vol. 96, No. 4 (August): 675–700.

Benedick, Richard E., 1991. *Ozone Diplomacy: New Directions in Safeguarding the Planet*. Cambridge, Mass.: Harvard University Press.

Berle, Adolf A., and Gardiner C. Means, 1932. *The Modern Corporation and Private Property*. New York: The MacMillan Company.

Bernstein, Paul M., W. David Montgomery, and Thomas F. Rutherford, 1999a. "Global Impacts of the Kyoto Agreement: Results from the MS-MRT Model," *Resource and Energy Economics*, Vol. 21: 375–413.

Bernstein, Paul M., W. David Montgomery, Thomas F. Rutherford, and Gui-Fang Yang, 1999b. "Effects of Restrictions on International Permit Trading," *The Energy Journal* (Special Issue), ed. John P. Weyant: 221–56.

Bobchuk, Lucian Ayre, Jesse M. Fried, and David I. Walker, 2001. "Executive Compensation in America: Optimal Contracting or Extraction of Rents," NBER Working Paper 8661. Cambridge, Mass.: National Bureau of Economic Research, http://www.nber.org/papers/w8661

Bohi, Douglas, and Dallas Burtraw, 1997. "SO_2 Allowance Trading: How Do Expectations and Experience Measure Up?" *The Electricity Journal* (August/September): 67–75.

Bollen, Johannes, Arjen Gielen, and Hans Timmer, 1999. "Clubs, Ceilings and CDM: Macroeconomics of Compliance with the Kyoto Protocol," *The Energy Journal* (Special Issue), ed. John P. Weyant: 177–206.

Bradley, Michael, Anand Desai, and E. Han Kim, 1988. "Synergistic Gains from Corporate Acquisitions and Their Division between the Stockholders of Target and Acquiring Firms," *Journal of Financial Economics*, Vol. 21, No. 1: 3–40.

Brekke, Kjell Arne, and Richard B. Howarth, 2000. "The Social Contingency of Wants," *Land Economics*, Vol. 76, No. 4: 493–503.

——, 2002. *Status, Growth and the Environment: Goods as Symbols in Applied Welfare Economics*. Cheltenham, UK, and Northampton, Mass.: Edward Elgar.

Broome, John, 1992. *Counting the Cost of Global Warming*. Cambridge, UK: The White Horse Press.

Brown, Donald A., 2002. *American Heat: Ethical Problems with the United States' Response to Global Warming*. Boulder, Colo.: Rowman & Littlefield Publishers, Inc.

Browner, Carol, 1997. Testimony before the Committee on Environment and Public Works, US Senate, February 12.

Button, Kenneth, ed., 1989. *The Collected Essays of Harvey Leibenstein*. Vol. 2. *X-efficiency and Micro-micro Theory*. Aldershot: Elgar.

Canan, Penelope, and Nancy Reichman, 2001. *Ozone Connections: Expert Networks in Global Environmental Governance*. Sheffield, UK: Greenleaf Publishing.

Carroll, Glenn R., and Michael T. Hannan, 2000. *The Demography of Corporations and Industries*. Princeton: Princeton University Press.

Carson, Richard T., Nicholas E. Flores, and Norman F. Meade, 2001. "Contingent Valuation: Controversies and Evidence," *Environmental and Resource Economics*, Vol. 19: 173–210.

Catholic Biblical Association of America, 1970. *The New American Bible*. Copyright owned by the Confraternity of Christian Doctrine, Washington, DC. New York: P.J. Kenedy & Sons.

Charnes, A., W.W. Cooper, A.Y. Lewin, and L.M. Seiford, 1994. *Data Envelopment Analysis: Theory, Methodology and Applications*. Norwell, Mass.: Kluwer Academic Publishers.

Chichilnisky, Graciela, 1997. "What Is Sustainable Development?" *Land Economics*, Vol. 73, No. 4 (November): 467–91.

Cline, William R., 1992. *The Economics of Global Warming*. Washington, DC: Institute for International Economics.

Coase, R.H., 1937. "On the Nature of the Firm," *Economica*, n.s. Vol. 4, No. 16: 386–405.

——, 1960. "The Problem of Social Cost," *Journal of Law and Economics*, Vol. 3, No. 1 (October): 1–44.

Conlisk, J., 1996. "Why Bounded Rationality?" *Journal of Economic Literature*, Vol. 34, No. 2: 669–700.

Cook, Elizabeth, ed., 1996. *Ozone Protection in the United States: Elements of Success*. Washington, DC: World Resources Institute.

Cooper, Adrian, Scott Livermore, Vanessa Rossi, Alan Wilson, and John Walker, 1999. "The Economic Implications of Reducing Carbon Emissions: a Cross-Country Quantitative Investigation using the Oxford Global Macroeconomic and Energy Model," *The Energy Journal* (Special Issue), ed. John P. Weyant: 335–65.

Cooper, William W., Lawrence M. Seiford, and Kaoru Tone, 2000. *Data Envelopment Analysis: a Comprehensive Text with Models, Applications, References and DEA-Solver Software*. Boston: Kluwer Academic Publishers.

Cooter, Robert, and Peter Rappoport, 1984. "Were the Ordinalists Wrong about Welfare Economics?" *Journal of Economic Literature*, Vol. 22 (June): 507–30.

Council of Economic Advisers [CEA], 1998a. *The Annual Report of the Council of Economic Advisers* [1998]. Washington, DC: US Government Printing Office.

——, 1998b. *The Kyoto Protocol and the President's Policies to Address Climate Change: Administration Economic Analysis*. Washington, DC: Council of Economic Advisers, July.

CPB Netherlands Bureau for Economic Policy Analysis, 1999. WorldScan: The Core Version, http://www.cpb.nl/eng/pub/bijzonder/20/bijz20.pdf

Crawford, Mark, 1987. "Ozone Plan Splits Administration," *Science*, New Series, Vol. 236, Issue 4805 (May 29): 1052–3.

Crutzen, P., and E. Stoermer, 2000. *International Geosphere Biosphere Programme (IGBP) Newsletter*, Vol. 41.

Dahlman, Carl J., 1979. "The Problem of Externality," *Journal of Law and Economics*, Vol. 22, No. 1 (April): 141–62.

Dales, J.H., 1968. *Pollution, Property and Prices*. Toronto: University of Toronto Press.

Debreu, G., 1970. "Economies with a Finite Set of Equilibria," *Econometrica*, Vol. 38: 387–92.

DeCanio, Samuel, 2000a. "Beyond Marxist State Theory: State Autonomy in Democratic Societies," *Critical Review*, Vol. 14, Nos 2–3: 215–36.

DeCanio, Stephen J., 1992. "Carbon Rights and Economic Development," *Critical Review*, Vol. 6, Nos 2–3: 389–410.

——, 1993. "Barriers within Firms to Energy-efficient Investments," *Energy Policy*, Vol. 21, No. 9 (September): 906–14.

——, 1994a. "Agency and Control Problems in US Corporations: the Case of Energy-efficient Investment Projects," *Journal of the Economics of Business*, Vol. 1, No. 1: 105–23.

——, 1994b. "Why Do Profitable Energy-Saving Investment Projects Languish?" *Journal of General Management*, Vol. 20, No. 1 (Autumn): 62–71.

——, 1994c. "Energy Efficiency and Managerial Performance: Improving Profitability While Reducing Greenhouse Gas Emissions," in *Global Climate Change and Public Policy*, ed. David L. Feldman. Chicago: Nelson-Hall Publishers.

——, 1997. "Economic Modeling and the False Tradeoff between Environmental Protection and Economic Growth," *Contemporary Economic Policy*, Vol. 15: 10–27.

——, 1998. "The Efficiency Paradox: Bureaucratic and Organizational Barriers to Profitable Energy-saving Investments," *Energy Policy*, Vol. 26, No. 5: 441–54.

——, 1999. "Estimating the Non-Environmental Consequences of Greenhouse Gas Reductions Is Harder Than You Think," *Contemporary Economic Policy*, Vol. 17, No. 3 (July): 279–95.

——, 2000b. "The Organizational Structure of Firms and Economic Models of Climate Policy," in Stephen J. DeCanio, Richard B. Howarth, Alan H. Sanstad, Stephen H. Schneider, and Starley L. Thompson, *New Directions in the Economics and Integrated Assessment of Global Climate Change*. Washington, DC: Pew Center on Global Climate Change.

——, 2002. "Economic Analysis and Environmental Policy in the Reagan Administration: the Case of the Montreal Protocol," paper presented at the Conference on the Reagan Presidency, University of California, Santa Barbara, March 27–30.

DeCanio, Stephen J., Catherine Dibble, and Keyvan Amir-Atefi, 2000. "The Importance of Organizational Structure for the Adoption of Innovations," *Management Science*, Vol. 46, No. 10 (October): 1285–99.

——, 2001a. "Organizational Structure and the Behavior of Firms: Implications for Integrated Assessment," *Climatic Change*, Vol. 48: 487–514.

DeCanio, Stephen J., and William E. Watkins, 1998. "Information Processing and Organizational Structure," *Journal of Economic Behavior and Organization*, Vol. 36, No. 3 (August): 275–94.

DeCanio, Stephen J., William E. Watkins, Glenn Mitchell, Keyvan Amir-Atefi, and Catherine Dibble, 2001b. "Complexity in Organizations: Consequences for Climate Policy Analysis," in *Advances in the Economics of Environmental Resources*, Vol. 3, *The Long-Term Economics of Climate Change: Beyond a Doubling of Greenhouse Gas Concentrations*, eds Darwin C. Hall and Richard B. Howarth. Amsterdam: JAI/Elsevier Science.

Desvousges, William H., F. Reed Johnson, Richard W. Dunford, Kevin J. Boyle, Sara P. Hudson, and K. Nicole Wilson, 1993. "Measuring Natural Resource

Damages with Contingent Valuation: Tests of Validity and Reliability," in *Contingent Valuation: a Critical Assessment*, ed. Jerry A. Hausman. Amsterdam: North-Holland.

Diamond, Peter A., and Jerry A. Hausman, 1994. "Contingent Valuation: Is Some Number Better than No Number?" *Journal of Economic Perspectives*, Vol. 8, No. 4 (Fall): 45–64.

Dierker, Egbert, 1972. "Two Remarks on the Number of Equilibria of an Economy," *Econometrica*, Vol. 40, No. 5 (September): 951–3.

Dockner, Engelbert J., and Gustav Feichtinger, 1993. "Cyclical Consumption Patterns and Rational Addiction," *The American Economic Review*, Vol. 83, No. 1 (March): 256–63.

Dosi, Giovanni, 2000. *Innovation, Organization and Economic Dynamics: Selected Essays*. Cheltenham, UK: Edward Elgar.

Ebert, Udo, 2001. "A General Approach to the Evaluation of Nonmarket Goods," *Resource and Energy Economics*, Vol. 23: 373–88.

Economist, The, 2001. "Sunset for the Oil Business?" November 3: 81–2.

Energy Modeling Forum [EMF], 1999. *The Costs of the Kyoto Protocol: a Multimodel Evaluation*. 16th Energy Modeling Forum, *The Energy Journal*, Special Issue, ed. John P. Weyant.

Farmer, Roger E.A., 1999. *Macroeconomics of Self-fulfilling Prophecies*. Cambridge, Mass.: The MIT Press.

Farrell, M.J., 1957. "The Measurement of Productive Efficiency," *Journal of the Royal Statistical Society*, Series A (General), Vol. 120, Part 3: 253–90.

Fellner, W., 1967. "Operational Utility: the Theoretical Background and a Measurement," in *Ten Economic Studies in the Tradition of Irving Fisher*. New York: John Wiley and Sons.

Fisher, Franklin M., 1965. "Embodied Technical Change and the Existence of an Aggregate Capital Stock," *Review of Economic Studies*, Vol. 32: 263–88.

——, 1966. *The Identification Problem in Econometrics*. New York: McGraw-Hill Book Company.

——, 1968a. "Embodied Technology and the Existence of Labour and Output Aggregates," *Review of Economic Studies*, Vol. 35: 391–412.

——, 1968b. "Embodied Technology and the Aggregation of Fixed and Movable Capital Goods," *Review of Economic Studies*, Vol. 35: 417–28.

——, 1969a. "Approximate Aggregation and the Leontief Conditions," *Econometrica*, Vol. 37: 457–69.

——, 1969b. "The Existence of Aggregate Production Functions," *Econometrica*, Vol. 37, No. 4 (October): 553–77.

——, 1983. *Disequilibrium Foundations of Equilibrium Economics*. Cambridge: Cambridge University Press.

——, 1984. "The Misuse of Accounting Rates of Return: Reply," *American Economic Review*, Vol. 74, No. 3 (June): 509–17.

——, 1989. "Adjustment Processes and Stability," in *The New Palgrave: General Equilibrium*, eds John Eatwell, Murray Milgate and Peter Newman, First American edition. New York and London: W.W. Norton.

Fisher, Franklin M., and John J. McGowan, 1983. "On the Misuse of Accounting Rates of Return to Infer Monopoly Profits," *American Economic Review*, Vol. 73 (March): 82–97.

Fisher, Franklin M., John J. McGowan, and Joen E. Greenwood, 1983. *Folded, Spindled, and Mutilated: Economic Analysis and U.S. v. IBM*. Cambridge: MIT Press.

Fontana, Walter, and Leo W. Buss, 1994. "What Would be Conserved if 'the Tape were Played Twice'?" *Proceedings of the National Academy of Sciences of the United States of America*, Vol. 91, No. 2 (January 18): 757–61.

Førsund, F.R., 1999. "The Evolution of DEA – the Economics Perspective," paper presented to the Sixth European Workshop on Efficiency and Productivity. Copenhagen, 29–31 October.

Frank, Robert H., 1985a. "The Demand for Unobservable and Other Nonpositional Goods," *American Economic Review*, Vol. 75, No. 1 (March): 101–16.

——, 1985b. *Choosing the Right Pond: Human Behavior and the Quest for Status*. New York: Oxford University Press.

Frederick, Shane, George Loewenstein, and Ted O'Donoghue, 2002. "Time Discounting and Time Preference: a Critical Review," *Journal of Economic Literature*, Vol. 40, No. 2 (June): 351–401.

Freixas, X., and A. Mas-Colell, 1987. "Engel Curves Leading to the Weak Axiom in the Aggregate," *Econometrica*, Vol. 55, No. 3 (May): 515–31.

Frey, Bruno S., and Alois Stutzer, 2002. "What Can Economists Learn from Happiness Research?" *Journal of Economic Literature*, Vol. 40, No. 2 (June): 402–35.

Friedman, Jeffrey, 1990. "The New Consensus: II. The Democratic Welfare State," *Critical Review*, Vol. 4, No. 4: 633–708.

——, 1997. "What's Wrong with Libertarianism," *Critical Review*, Vol. 11, No. 3: 407–67.

Friedman, Milton, 1953. *Essays in Positive Economics*. Chicago: University of Chicago Press.

Garey, Michael R., and David S. Johnson, 1979 [updated 1991]. *Computers and Intractability: a Guide to the Theory of NP-Completeness*. New York: Freeman.

Geanakoplos, John, 1989a. "Arrow–Debreu Model of General Equilibrium," in *The New Palgrave: General Equilibrium*, eds John Eatwell, Murray Milgate, and Peter Newman. New York: W.W. Norton & Company, Inc.

——, 1989b. "Overlapping Generations Model of General Equilibrium," in *The New Palgrave: General Equilibrium*, eds John Eatwell, Murray Milgate, and Peter Newman. New York: W.W. Norton & Company, Inc.

Gerlagh, R., and B.C.C. van der Zwaan, 2001a. "Overlapping Generations versus Infinitely-Lived Agent: the Case of Global Warming," in *Advances in the Economics of Environmental Resources*, Vol. 3, *The Long-Term Economics of Climate Change: Beyond a Doubling of Greenhouse Gas Concentrations*, eds Darwin C. Hall and Richard B. Howarth. Amsterdam: JAI/Elsevier Science.

——, 2001b. "The Effects of Ageing and an Environmental Trust Fund in an Overlapping Generations Model on Carbon Emission Reductions," *Ecological Economics*, Vol. 36: 311–26.

Goldberg, David E., 1989. *Genetic Algorithms in Search, Optimization, and Machine Learning*. Menlo Park, Calif.: Addison-Wesley Publishing Company, Inc.

Goodstein, Eban, 1997. "Polluted Data," *The American Prospect*, Vol. 8, No. 35.

Gould, S.J., 1989. *This Wonderful Life*. New York: Norton.

Goulder, L.H., and S.H. Schneider, 1999. "Induced Technological Change and the Attractiveness of CO_2 Abatement Policies," *Resource and Energy Economics*, Vol. 21, Nos 3–4 (August): 211–53.

Haddad, Brent M. and Richard B. Howarth, forthcoming. "Protest Bids, Commensurability, and Substitution: Contingent Valuation and Ecological Economics," in *Handbook of Contingent Valuation*, eds J.R. Kahn, A. Alberini, and D. Bjornstad. Cheltenham: Edward Elgar.

Hahn, F.H., 1989. "Auctioneer," in *The New Palgrave: General Equilibrium*, eds John Eatwell, Murray Milgate and Peter Newman, First American edition. New York and London: W.W. Norton.

Hall, Robert E., 1988. "Intertemporal Substitution in Consumption," *Journal of Political Economy*, Vol. 96, No. 2 (April): 339–57.

Hammitt, James K., 1997. "Are the Costs of Proposed Environmental Regulations Overestimated? Evidence from the CFC Phaseout." Center for Risk Analysis and Department of Health Policy and Management, Harvard School of Public Health. Cambridge, Mass. (May).

Hanemann, W. Michael, 1994. "Valuing the Environment through Contingent Valuation," *Journal of Economic Perspectives*, Vol. 8, No. 4 (Fall): 19–43.

Hannan, Michael T., and John Freeman, 1989. *Organizational Ecology*. Cambridge: Harvard University Press.

Hanoch, Giora, 1975. "Production and Demand Models with Direct or Indirect Implicit Additivity," *Econometrica*, Vol. 43, No. 3 (May): 395–420.

Harary, Frank, 1969. *Graph Theory*. Reading, Mass.: Addison-Wesley Publishing Company.

Harcourt, G.C., 1965. "The Accountant in a Golden Age," *Oxford Economic Papers*, Vol. 17 (March): 66–80.

Härdle, Wolfgang, Werner Hildenbrand, and Michael Jerison, 1991. "Empirical Evidence on the Law of Demand," *Econometrica*, Vol. 59, No. 6 (November): 1525–49.

Harrington, Winston, Richard D. Morgenstern, and Peter Nelson, 2000. "On the Accuracy of Regulatory Cost Estimates," *Journal of Policy Analysis and Management*, Vol. 19, No. 2: 297–322.

Harris, Ellie G., 1990. "Antitakeover Measures, Golden Parachutes, and Target Firm Shareholder Welfare," *The RAND Journal of Economics*, Vol. 21, No. 4 (Winter): 614–25.

Hausman, Jerry A., ed., 1993. *Contingent Valuation: a Critical Assessment*. (Contributions to Economic Analysis: 220). Amsterdam: North-Holland.

Hayek, Frederich A. von, 1988. *The Fatal Conceit: the Errors of Socialism*, in *The Collected Works of F.A. Hayek*, Vol. 1, ed. W.W. Bartley III. Chicago: University of Chicago Press.

Heinzerling, Lisa, and Frank Ackerman, 2002. "Pricing the Priceless: Cost–Benefit Analysis of Environmental Protection," Georgetown University: Georgetown Environmental Law and Policy Institute and Georgetown University Law Center.

Herings, Jean-Jacques, 1998. "Appendix: Recent Developments in Computing Equilibrium Prices," in *Elements of General Equilibrium Analysis*, ed. Alan Kirman. Oxford: Blackwell Publishers, Ltd.

Hildenbrand, Werner, 1983. "On the 'Law of Demand,'" *Econometrica*, Vol. 51, No. 4 (July): 997–1020.

——, 1994. *Market Demand: Theory and Empirical Evidence*. Princeton, NJ: Princeton University Press.

Hodgson, Geoffrey M., 1993. *Economics and Evolution, Bringing Life Back into Economics*. Ann Arbor: University of Michigan Press.

Hoffert, Martin I., Ken Caldeira, Atul K. Jain, Erik F. Haites, L.D. Danny Harvey, Seth D. Potter, Michael E. Schlesinger, Stephen H. Schneider, Robert G. Watts, Tom M.L. Wigley, and Donald J. Wuebbles, 1998. "Energy Implications of Future Stabilization of Atmospheric CO_2 Content," *Nature*, Vol. 395 (29 October): 881–4.

Hoffert, Martin I., Ken Caldeira, Gregory Benford, David R. Criswell, Christopher Green, Howard Herzog, Atul K. Jain, Haroon S. Kheshgi, Klaus S. Lackner, John S. Lewis, H. Douglas Lightfoot, Wallace Manheimer, John C. Mankins, Michael E. Mauel, L. John Perkins, Michael E. Schlesinger, Tyler Volk, and Tom M.L. Wigley, 2002. "Advanced Technology Paths to Global Climate Stability: Energy for a Greenhouse Planet," *Science*, Vol. 298 (1 November): 981–7.

Holland, J.H., 1975. *Adaptation in Natural and Artificial Systems*. Ann Arbor: University of Michigan Press.

Holte, Susan H., 2000. "Annual Energy Outlook Forecast Evaluation," US Department of Energy, Energy Information Administration, http://www.eia.doe.gov/oiaf/analysispaper/forecast_eval.html

Horowitz, Ira, 1984. "The Misuse of Accounting Rates of Return: Comment," *American Economic Review*, Vol. 74, No. 3 (June): 492–3.

Howarth, Richard B., 1996. "Status Effects and Environmental Externalities," *Ecological Economics*, Vol. 16, No. 1: 25–34.

——, 2000. "Climate Change and Relative Consumption," in *Advances in Global Change Research*, Vol. 8, *Society, Behaviour, and Climate Change Mitigation*, eds Eberhard Jochem, Jayant Sathaye, and Daniel Bouille. Dordrecht: Kluwer Academic Publishers.

Howarth, R.B., and R.B. Norgaard, 1992. "Environmental Valuation under Sustainable Development," *American Economic Review*, Vol. 82: 473–7.

Ingrao, Bruna, and Giorgio Israel, 1990. *The Invisible Hand: Economic Equilibrium in the History of Science*. Cambridge, Mass.: MIT Press.

IPCC [Intergovernmental Panel on Climate Change], 1996. *Climate Change 1995: Economic and Social Dimensions of Climate Change, Contribution of Working Group III to the Second Assessment Report of the Intergovernmental Panel on Climate Change*, eds James P. Bruce, Hoesung Lee, and Erik F. Haites. Cambridge: Cambridge University Press.

——, 2001a. *Climate Change 2001: the Scientific Basis. Contribution of Working Group I to the Third Assessment Report of the Intergovernmental Panel on Climate Change*, eds Houghton, J.T., Y Ding, D.J. Griggs, M. Noguer, P.J. van der Linden, X. Dai, K. Maskell, and C.A. Johnson. Cambridge, UK, and New York: Cambridge University Press.

——, 2001b. *Climate Change 2001: Synthesis Report. A Contribution of Working Groups I, II, and III to the Third Assessment Report of the Intergovernmental Panel on Climate Change*, eds R.T. Watson and the Core Writing Team. Cambridge, UK, and New York: Cambridge University Press.

Interlaboratory Working Group [IWG], 1997. *Scenarios of US Carbon Reductions: Potential Impacts of Energy-Efficient and Low-Carbon Technologies by 2010 and Beyond*, Lawrence Berkeley National Laboratory, Berkeley, Calif., and Oak Ridge National Laboratory, Oak Ridge, Tenn., September, http://www.ornl.gov/ORNL/Energy_Eff/labweb.html

——, 2000. *Scenarios for a Clean Energy Future*. Interlaboratory Working Group on Energy-Efficient and Clean-Energy Technologies, Oak Ridge National Laboratory and Lawrence Berkeley National Laboratory. ORNL/CON-476 and LBNL-44029, Oak Ridge, Tenn. and Berkeley, Calif., November, http://www.ornl.gov/ORNL/Energy_Eff/CEF.htm

ISI Web of Science, 2002. Social Sciences Citation Index. Philadelphia: Thomson/ISI. http://www.isinet.com/isi/products/citation/wos/

Jacobs, Michael, 1994. "The Limits to Neoclassicism: Towards an Institutional Environmental Economics," in *Social Theory and the Global Environment*, eds Michael Redclift and Ted Benton. London: Routledge.

Jacoby, Henry D., and Ian Sue Wing, 1999. "Adjustment Time, Capital Malleability, and Policy Cost," *The Energy Journal* (Special Issue), ed. John P. Weyant: 73–92.

Jarrell, Gregg A., James A. Brickley, and Jeffry M. Netter, 1988. "The Market for Corporate Control: the Empirical Evidence since 1980," *Journal of Economic Perspectives*, Vol. 2, No. 1: 49–68.

Jensen, Michael C., 1988. "Takeovers: Their Causes and Consequences," *Journal of Economic Perspectives*, Vol. 2, No. 1: 21–48.

Jensen, Michael C., and William H. Meckling, 1976. "Theory of the Firm: Managerial Behavior, Agency Costs and Ownership Structure," *Journal of Financial Economics*, Vol. 3, No. 4 (October): 305–60.

Jerison, M., 1984. "The Representative Consumer and the Weak Axiom when the Distribution of Income Is Fixed," working paper, Department of Economics, SUNY Albany.

Jochem, Eberhard, Jayant Sathaye, and Daniel Bouille, eds, 2000. *Advances in Global Change Research*, Vol. 8, *Society, Behaviour, and Climate Change Mitigation*. Dordrecht: Kluwer Academic Publishers.

Kainuma, Mikiko, Yuzuru Matsuoka, and Tsuneyuki Morita, 1999. "Analysis of Post-Kyoto Scenarios: the Asian-Pacific Integrated Model," *The Energy Journal* (Special Issue), ed. John P. Weyant: 207–20.

Kandel, Shmuel, and Robert F. Stambaugh, 1991. "Asset Returns and Intertemporal Preferences," *Journal of Monetary Economics*, Vol. 27, No. 1 (February): 39–71.

Karpoff, Jonathan M., and Paul H. Malatesta, 1989. "The Wealth Effects of Second-Generation State Takeover Legislation," *Journal of Financial Economics*, Vol. 25: 291–322.

Kaya, Y., 1989. "Impact of Carbon Dioxide Emission Control on GNP Growth: Interpretation of Proposed Scenarios," Intergovernmental Panel on Climate Change/Response Strategies Working Group (May), Geneva.

Kehoe, Timothy J., 1985. "Multiplicity of Equilibria and Comparative Statics," *Quarterly Journal of Economics*, Vol. 100, No. 1 (February): 119–47.

——, 1998. "Uniqueness and Stability," in *Elements of General Equilibrium Analysis*, ed. Alan Kirman. Oxford: Blackwell Publishers, Ltd.

Kehoe, Timothy J., and David K. Levine, 1990. "The Economics of Indeterminacy in Overlapping Generations Models," *Journal of Public Economics*, Vol. 42: 219–43.

Kehoe, Timothy J., and John Whalley, 1985. "Uniqueness of Equilibrium in Large-Scale Numerical General Equilibrium Models," *Journal of Public Economics*, Vol. 28: 247–54.

Kirman, Alan P., 1992. "Whom or What Does the Representative Individual Represent?" *Journal of Economic Perspectives*, Vol. 6, No. 2 (Spring): 117–36.

Kocherlakota, N.R., 1996. "The Equity Premium: It's Still a Puzzle," *Journal of Economic Literature*, Vol. 34: 42–71.

Koomey, Jonathan G., Paul Craig, and Ashok Gadgil, 2001. "Looking Ahead: the Perils and Promise of Long-term Forecasts," unpublished manuscript.

Koopmans, T.C., 1951. "Analysis of Production as an Efficient Combination of Activities," in *Activity Analysis of Production and Allocation*, ed. T.C. Koopmans. New York: Wiley.

——, 1960. "Stationary Ordinal Utility and Impatience," *Econometrica*, Vol. 28: 287–309.

——, 1965. "On the Concept of Optimal Economic Growth," in *The Econometric Approach to Development Planning*. Amsterdam: North-Holland, and Chicago: Rand McNally.

Kopp, R., R. Morgenstern, and W. Pizer, 1997. "Something for Everyone: a Climate Policy that Both Environmentalists and Industry Can Live With." Washington, DC: Resources for the Future, http://www.weathervane.rff.org/features/feature015.html

Krause, Florentin, in collaboration with Paul Baer and Stephen J. DeCanio, 2001. "Cutting Carbon Emissions at a Profit: Executive Summary." El Cerrito, Calif.: International Project for Sustainable Energy Paths.

Krause, Florentin, Stephen J. DeCanio, J. Andrew Hoerner, and Paul Baer, 2002. "Cutting Carbon Emissions at a Profit (Part I): Opportunities for the U.S." *Contemporary Economic Policy*, Vol. 20, No. 4 (October): 339–65.

——, 2003. "Cutting Carbon Emissions at a Profit (Part II): Impacts on U.S. Competitiveness and Jobs," *Contemporary Economic Policy*, Vol. 21, No. 1 (January): 90–105.

Krugman, Paul, 1995. *Peddling Prosperity: Economic Sense and Nonsense in an Age of Diminished Expectations*. New York: W.W. Norton & Company.

Kurosawa, Atsushi, Hiroshi Yagita, Weisheng Zhou, Koji Tokimatsu, and Yukio Yanagisawa, 1999. "Analysis of Carbon Emission Stabilization Targets and Adaptation by Integrated Assessment Model," *The Energy Journal* (Special Issue), ed. John P. Weyant: 157–75.

Laitner, John A. "Skip," Stephen J. DeCanio, and Irene Peters, 2000. "Incorporating Behavioral, Social, and Organizational Phenomena in the Assessment of Climate Change Mitigation Options," in *Society, Behaviour, and Climate Change Mitigation*, eds E. Jochem, D. Bouille, and J. Sathaye. New York: Kluwer Academic Press.

Lane, Robert E., 2001. *The Loss of Happiness in Market Democracies*. New Haven: Yale University Press.

Leibenstein, Harvey, 1966. "Allocative Efficiency vs. 'X-efficiency,'" *American Economic Review*, Vol. 56, No. 3 (June): 392–415.

Leibenstein, Harvey, and Shlomo Maital, 1992. "Empirical Estimation and Partitioning of X-Inefficiency: a Data-Envelopment Approach," *American Economic Review*, Vol. 82, No. 2 (May): 428–33.

——, 1994. "The Organizational Foundations of X-Inefficiency: a Game-Theoretic Interpretation of Argyris' Model of Organizational Learning," *Journal of Economic Behavior and Organization*, Vol. 23, No. 3 (May): 251–68.

188   *References*

Loewenstein, George, 2000. "Emotions in Economic Theory and Economic Behavior," *American Economic Review: Papers and Proceedings*, Vol. 90, No. 2 (May): 426–32.

Long, William F., and David J. Ravenscraft, 1984. "The Usefulness of Accounting Profit Data: a Comment on Fisher and McGowan," *American Economic Review*, Vol. 74, No. 3 (June): 494–500.

Lovins, Amory, 1976. "Energy Strategy: the Road Not Taken?" *Foreign Affairs; an American Quarterly Review*, Vol. 55, No. 1 (October): 65–96.

Lucas, Robert E., Jr, 1988. "On the Mechanics of Economic Development," *Journal of Monetary Economics*, Vol. 22 (July): 3–42.

Macaulay, Hugh H., and Bruce Yandle, 1977. *Environmental Use and the Market*. Lexington, Mass.: Lexington Books.

MacCracken, Christopher N., James A. Edmonds, Son H. Kim, and Ronald D. Sands, 1999. "The Economics of the Kyoto Protocol," *The Energy Journal* (Special Issue), ed. John P. Weyant: 25–71.

McKibbin, Warwick J., 2000. "Moving beyond Kyoto," Brookings Institution Policy Brief No. 66. Washington, DC: The Brookings Institution.

McKibbin, Warwick J., Martin T. Ross, Robert Shackleton, and Peter J. Wilcoxen, 1999. "Emissions Trading, Capital Flows, and the Kyoto Protocol," *The Energy Journal* (Special Issue), ed. John P. Weyant: 287–333.

McKibbin, W., and P. Wilcoxen, 1997a. "A Better Way to Slow Global Climate Change," Brookings Policy Brief no. 17. Washington, DC: The Brookings Institution, http://www.brook.edu/comm/PolicyBriefs/pb017/ pb17.htm

——, 1997b. "Salvaging the Kyoto Climate Change Negotiations," Brookings Policy Brief no. 27. Washington, DC: The Brookings Institution, http://www.brook.edu/comm/PolicyBriefs/pb027/pb27.htm

Manne, Alan S., and Richard G. Richels, 1999. "The Kyoto Protocol: a Cost-Effective Strategy for Meeting Environmental Objectives?" *The Energy Journal* (Special Issue), ed. John P. Weyant: 1–23.

Martin, Stephen, 1984. "The Misuse of Accounting Rates of Return: Comment," *American Economic Review*, Vol. 74, No. 3 (June): 501–6.

Marwell, G., and R. Ames, 1981. "Economists Free Ride: Does Anyone Else?" *Journal of Public Economics*, Vol. 15: 295–310.

Mas-Colell, Andreu, Michael D. Whinston, and Jerry R. Green, 1995. *Microeconomic Theory*. New York: Oxford University Press.

Meadows, Donella H., Dennis L. Meadows, Jørgen Randers, and William W. Behrens III, 1972. *The Limits to Growth: a Report for the Club of Rome's Project on the Predicament of Mankind*. New York: Universe Books.

Mehra, Rajnish, and Edward C. Prescott, 1985. "The Equity Premium: a Puzzle," *Journal of Monetary Economics*, Vol. 15, No. 2: 145–61.

Mérö, László, 1998. *Moral Calculations: Game Theory, Logic, and Human Frailty*, translated by Anna C. Gösi-Greguss, English version edited by David Kramer. New York: Springer-Verlag.

Metz, Bert, et al., 2000. *Methodological and Technological Issues in Technology Transfer*, Special Report of IPCC Working Group III, Intergovernmental Panel on Climate Change. Cambridge: Cambridge University Press.

Mitchell, Glenn T., 2001. "Evolution of Networks and the Diffusion of New Technology," Chapter 1 of Ph.D. dissertation, "Technology, Energy and the Environment," University of California, Santa Barbara.

Mitchell, Melanie, 1996. *An Introduction to Genetic Algorithms (Complex Adaptive Systems)*. Cambridge, Mass.: MIT Press.

Moore, Randell E., ed., 2002. *Blue Chip Economic Indicators*, Vol. 27, No. 3 (March 10).

Mooz, W.E., S.H. Dole, D.L. Jaquette, W.H. Krase, P.F. Morrison, S.L. Salem, R.G. Salter, and K.A. Wolf, 1982. *Technical Options for Reducing Chlorofluorocarbon Emissions*, prepared for the US Environmental Protection Agency. Santa Monica, Calif.: The RAND Corporation.

Nash, Hugh, ed., 1979. *The Energy Controversy: Soft Path Questions & Answers*. San Francisco: Friends of the Earth.

NBER, 2002. "Behavioral Finance," http://www.nber.org/reporter/summer02/news/bf.html

Negishi, Takashi, 1960. "Welfare Economics and Existence of an Equilibrium for a Competitive Economy," *Metroeconomica*, Vol. 12: 92–7.

Nelson, Richard R., 1995. "Recent Evolutionary Theorizing about Economic Change," *Journal of Economic Literature*, Vol. 33, No. 1 (March): 48–90.

Nelson, Richard R., and Sidney G. Winter, 1982. *An Evolutionary Theory of Economic Change*. Cambridge: The Belknap Press of Harvard University Press.

Nerlove, Marc, 1958. *The Dynamics of Supply: Estimation of Farmers' Response to Price*. Baltimore: Johns Hopkins Press.

Neumayer, Eric, 1999. "Global Warming: Discounting is not the Issue, but Substitutability is," *Energy Policy*, Vol. 27: 33–43.

Newbold, Paul, 1995. *Statistics for Business & Economics*, 4th edition. Englewood Cliffs, NJ: Prentice-Hall.

NRDC, 2002. "Landmark Power Plant Clean-Up Bill Headed to Senate Floor; Provides Pollution Relief, Counters White House Assault on Clean Air Safeguards," press release, Natural Resources Defense Council, http://www.nrdc.org/media/pressreleases/020627.asp

Nordhaus, William D., 1979. *The Efficient Use of Energy Resources*. New Haven: Yale University Press.

——, 1994. *Managing the Global Commons: the Economics of Climate Change*. Cambridge, Mass.: The MIT Press.

Nordhaus, William D., and Joseph G. Boyer, 1999. "Requiem for Kyoto: an Economic Analysis of the Kyoto Protocol," *The Energy Journal* (Special Issue), ed. John P. Weyant: 93–130.

——, 2000. *Warming the World: Economic Models of Global Warming*. Cambridge: MIT Press.

Obstfeld, Maurice, and Kenneth Rogoff, 1996. *Foundations of International Macroeconomics*. Cambridge, Mass.: The MIT Press.

Olson, Mancur, Jr, 1965. *The Logic of Collective Action: Public Goods and the Theory of Groups*. Cambridge, Mass.: Harvard University Press.

O'Neill, Brian C., and Michael Oppenheimer, 2002. "Dangerous Climate Impacts and the Kyoto Protocol," *Science*, Vol. 296 (June): 1971–2.

Palmer, Adele R. William E. Mooz, Timothy H. Quinn, and Kathleen A. Wolf, 1980. *Economic Implications of Regulating Chlorofluorocarbon Emissions from Non-aerosol Applications*, prepared for the US Environmental Protection Agency. Santa Monica, Calif.: The RAND Corporation.

Palmer, Karen, Wallace E. Oates, and Paul Portney, 1995. "Tightening Environmental Standards: the Benefit-Cost or the No-Cost Paradigm?" *Journal of Economic Perspectives*, Vol. 9, No. 4 (Fall): 119–32.

Papadimitriou, Christos H., 1996. "Computational Aspects of Organization Theory (Extended Abstract)," in *Algorithms – ESA '96: Proceedings Fourth Annual European Symposium*, Barcelona, Spain, September, eds Josep Diaz and Maria Serna. New York: Springer.

Pearce, D.W., and D. Ulph, 1994. "Estimating a Social Discount Rate for the United Kingdom," mimeo, Centre for Social and Economic Research on the Global Environment, University College London and University of East Anglia.

Peck, Stephen C., and Thomas J. Teisberg, 1992. "CETA: a Model for Carbon Emissions Trajectory Assessment," *The Energy Journal*, Vol. 13, No. 1: 55–77.

——, 1999. "CO$_2$ Emissions Control Agreements: Incentives for Regional Participation," *The Energy Journal* (Special Issue), ed. John P. Weyant: 367–90.

Pinches, George E., 1982. "Myopia, Capital Budgeting, and Decision Making," *Financial Management* (Autumn): 6–19.

Popper, Karl R., 1968 [1934]. *The Logic of Scientific Discovery* [*Logik der Forschung*]. New York: Harper & Row, Publishers, Incorporated.

Porter, Michael E., 1991. "America's Green Strategy: Environmental Standards and Competitiveness," *Scientific American*, Vol. 264, No. 4 (April): 168.

Porter, Michael E., and Claas van der Linde, 1995a. "Toward a New Conception of the Environment–Competitiveness Relationship," *Journal of Economic Perspectives*, Vol. 9, No. 4 (Fall): 97–118.

——, 1995b. "Green and Competitive: Breaking the Stalemate," *Harvard Business Review*, Vol. 73, No. 5 (September–October): 120–34.

Portney, Paul R., 1994. "The Contingent Valuation Debate: Why Economists Should Care," *Journal of Economic Perspectives*, Vol. 8, No. 4 (Fall): 3–17.

Pugh, William N., and John S. Jahera, Jr, 1990. "State Antitakeover Legislation and Shareholder Wealth," *Journal of Financial Research*, Vol. 13, No. 3: 221–31.

Quantitative Micro Software, 1997. *EViews 3 User's Guide*. Irvine, Calif.: Quantitative Micro Software.

Ramsey, F.P., 1928. "A Mathematical Theory of Saving," *Economic Journal*, Vol. 38: 543–59.

Reichman, Nancy, Penelope Canan, Stephen J. DeCanio, and Catherine Dibble, 2001. "Individual Leadership Matters: the Case of Global Ozone Layer Protection," unpublished manuscript.

Repetto, Robert, 2001. *Yes, Virginia, There is a Double Dividend*. Denver: Institute for Policy Research and Implementation, University of Colorado at Denver.

Repetto, Robert, and Duncan Austin, 1997. *The Costs of Climate Protection: a Guide for the Perplexed*. Washington, DC: World Resources Institute.

Riker, William H., and Peter C. Ordeshook, 1973. *An Introduction to Positive Political Theory*. Englewood Cliffs, NJ: Prentice-Hall, Inc.

Rizvi, S. Abu Turab, 1994. "The Microfoundations Project in General Equilibrium Theory," *Cambridge Journal of Economics*, Vol. 18: 357–77.

Robinson, Joan, 1953. "The Production Function and the Theory of Capital," *Review of Economic Studies*, Vol. 21, No. 2 (1953–54): 81–106.

Robinson, John B., 1988. "Unlearning and Backcasting – Rethinking Some of the Questions We Ask about the Future," *Technological Forecasting and Social Change*, Vol. 33, No. 4 (July): 325–38.

Romer, Paul M., 2000. "Thinking and Feeling," *American Economic Review: Papers and Proceedings*, Vol. 90, No. 2 (May): 439–43.

Rosenberg, Nathan, and L.E. Birdzell, Jr., 1986. *How the West Grew Rich: the Economic Transformation of the Industrial World*. New York: Basic Books, Inc.

Rust, John, 1996 [revised 1997]. "Dealing with the Complexity of Economic Calculations," paper for "Fundamental Limits to Knowledge in Economics," Workshop, Santa Fe Institute, 3 August.

Samuelson, Paul A., 1937. "A Note on Measurement of Utility," *Review of Economic Studies*, Vol. 4: 155–61.

——, 1958. "An Exact Consumption Loan Model of Interest, with or without the Social Contrivance of Money," *Journal of Political Economy*, Vol. 66, No. 5: 467–82.

Sands, Ronald D., James A. Edmonds, and Christopher N. MacCracken, 1999. "SGM 2000: Model Description and Theory," Pacific Northwest National Laboratory (September 17), manuscript.

Sanstad, Alan H., 2000. "Endogenous Technological Change and Climate Policy Modeling," in Stephen J. DeCanio, Richard B. Howarth, Alan H. Sanstad, Stephen H. Schneider, and Starley L. Thompson, *New Directions in the Economics and Integrated Assessment of Global Climate Change*. Washington: Pew Center on Global Climate Change.

Sanstad, Alan H., Stephen J. DeCanio, Gale A. Boyd and Jonathan G. Koomey, 2001a. "Estimating Bounds on the Economy-wide Effects of the CEF Policy Scenarios," *Energy Policy*, Vol. 29, No. 14 (November): 1299–311.

Sanstad, Alan H., Jonathan G. Koomey, and John A. "Skip" Laitner, 2001b. "Back to the Future: Long-range U.S. Energy Price and Quantity Projections in Retrospect," unpublished manuscript.

Schranz, Mary S., 1993. "Takeovers Improve Firm Performance: Evidence from the Banking Industry," *Journal of Political Economy*, Vol. 101, No. 2: 299–326.

Scott, Maurice FitzGerald, 1989. *A New View of Economic Growth*. Oxford: Clarendon Press.

Seiford, Lawrence M., 2000. "A Cyber-Bibliography for Data Envelopment Analysis (1978–1999) (September, 1999)," in William W. Cooper, Lawrence M. Seiford, and Kaoru Tone, *Data Envelopment Analysis: a Comprehensive Text with Models, Applications, References and DEA-Solver Software*. Boston: Kluwer Academic Publishers.

Sen, Amartya K., 1979. "The Welfare Basis of Real Income Comparisons: a Survey," *Journal of Economic Literature*, Vol. 17, No. 1: 1–45.

Sengupta, Jati K., 1995. *Dynamics of Data Envelopment Analysis: Theory of Systems Efficiency*. Dordrecht: Kluwer Academic Publishers.

Shlyakhter, Alexander I., Daniel M. Kammen, Claire L. Broido, and Richard Wilson, 1994. "Quantifying the Credibility of Energy Projections from Trends in Past Data," *Energy Policy*, Vol. 22, No. 2 (February): 119–30.

Shor, Peter W., 1997. "Polynomial-Time Algorithms for Prime Factorization and Discrete Logarithms on a Quantum Computer," *SIAM Journal of Computing*, Vol. 26: 1484–509.

Shoven, John B., and John Whalley, 1998 [1992]. *Applying General Equilibrium*. Cambridge: Cambridge University Press.

Shubik, Martin, 1971. "The Dollar Auction Game: a Paradox in Non-cooperative Behavior and Escalation," *Journal of Conflict Resolution*, Vol. 15: 109–11.

Siegel, Sidney, 1956. *Nonparametric Statistics for the Behavioral Sciences*. New York: McGraw-Hill Book Company, Inc.

Silberberg, Eugene, 1990. *The Structure of Economics: a Mathematical Analysis*. New York: McGraw-Hill Publishing Company.

Simon, Herbert A., 1963. "Discussion," *American Economic Review*, Vol. 53, No. 2: 229–31.

Solow, R.M., 1957. "Technical Change and the Aggregate Production Function," *The Review of Economics and Statistics*, Vol. 39: 312–20.

——, 1970. *Growth Theory: an Exposition*. New York: Oxford University Press.

Spear, Stephen E., 1989. "Learning Rational Expectations under Computability Constraints," *Econometrica*, Vol. 57, No. 4: 889–910.

Statman, Meir, and James F. Sepe, 1984. "Managerial Incentive Plans and the Use of the Payback Method," *Journal of Business Finance & Accounting*, Vol. 11, No. 1 (Spring): 61–5.

Tobin, James, 1986. "The Future of Keynesian Economics," *Eastern Economic Journal*, Vol. 13, No. 4.

Tol, R.S.J., 1994. "The Damage Costs of Climate Change – a Note on Tangibles and Intangibles, Applied to DICE," *Energy Policy*, Vol. 22, No. 5 (May): 436–8.

——, 1999. "Kyoto, Efficiency, and Cost-Effectiveness: Applications of FUND," *The Energy Journal* (Special Issue), ed. John P. Weyant: 131–56.

——, 2001. "Utility in FUND," personal communication (January 27).

Tulpulé, Vivek, Stephen Brown, Jaekyu Lim, Cain Polidano, Hom Pant, and Brian S. Fisher, 1999. "The Kyoto Protocol: an Economic Analysis Using GTEM," *The Energy Journal* (Special Issue), ed. John P. Weyant: 257–85.

US Bureau of Mines, 1970. *Mineral Facts and Problems, 1970*. Washington, DC: Government Printing Office.

US Congress, Office of Technology Assessment, 1995. *Gauging Control Technology and Regulatory Impacts in Occupational Safety and Health: an Appraisal of OSHA's Analytic Approach*. Washington, DC.

US Council of Economic Advisers, 1998a. *Economic Report of the President*. Washington, DC.

——, 1998b. *The Kyoto Protocol and the President's Policies to Address Climate Change: Administration Economic Analysis*. Washington, DC (July).

US Department of Commerce, Bureau of Economic Analysis, 2001. "Gross Domestic Product: Implicit Price Deflator," http://www.stls.frb.org/fred/data/gdp/gdpdef

US Department of Energy, 1987. *Energy Security: a Report to the President of the United States*. Washington, DC (March).

US Energy Information Administration [EIA], 1994. *Historical Monthly Energy Review, 1973–1992*. US Department of Energy, Office of Energy Markets and End Use. Washington, DC (August).

——, 1997. *Annual Energy Outlook 1998, with Projections to 2020*. US Department of Energy, Office of Integrated Analysis and Forecasting. Washington, DC (December).

——, 1998a. *Annual Energy Review 1997*. US Department of Energy, Office of Energy Markets and End Use. Washington, DC.

——, 1998b. *Impacts of the Kyoto Protocol on Energy Markets and Economic Activity*. US Department of Energy, Energy Information Administration, Office of Integrated Analysis and Forecasting. Washington, DC (October).

——, 2000a. *Annual Energy Outlook 2001, with Projections to 2020*. US Department of Energy, Office of Integrated Analysis and Forecasting. Washington, DC (December).

——, 2000b. *Analysis of Strategies for Reducing Multiple Emissions from Power Plants: Sulfur Dioxide, Nitrogen Oxides, and Carbon Dioxide*, SR/OIAF/2000-05. Washington, DC (December).

——, 2001a. *Analysis of Strategies for Reducing Multiple Emissions from Power Plants: Sulfur Dioxide, Nitrogen Oxides, Carbon Dioxide, and Mercury and a Renewable Portfolio Standard*, SR/OIAF/2001-03. Washington, DC (July).

——, 2001b. *Analysis of Strategies for Reducing Multiple Emissions from Electric Power Plants with Advanced Technology Scenarios*, SR/OIAF/2001-05. Washington, DC (October).

——, 2001c. *Annual Energy Outlook 2002 with Projections to 2020*. US Department of Energy, Office of Integrated Analysis and Forecasting. Washington, DC (December).

Urzúa, Carlos M., 1996. "On the Correct Use of Omnibus Tests for Normality," *Economics Letters*, Vol. 53: 247–51.

Van Breda, Michael F., 1984. "The Misuse of Accounting Rates of Return: Comment," *American Economic Review*, Vol. 74, No. 3 (June): 507–8.

Viscusi, W. Kip, 1991. "Age Variations in Risk Perceptions and Smoking Decisions," *The Review of Economics and Statistics*, Vol. 73, No. 4 (November): 577–88.

——, 1993. "The Value of Risks to Life and Health," *Journal of Economic Literature*, Vol. 31, No. 4 (December): 1912–46.

Weibull, Jörgen W., 1995. *Evolutionary Game Theory*. Cambridge, Mass.: The MIT Press.

Weinberg, D.H., 1996. "A Brief Look at Postwar US Income Inequality." *Current Population Reports*. US Census Bureau P60-191,
http://www.census.gov/hhes/www/p60191.html

Weyant, John P., 2000. *An Introduction to the Economics of Climate Change Policy*. Washington, DC: Pew Center on Global Climate Change.

Weyant, John P., and J.N. Hill, 1999. "Introduction and Overview," in *The Costs of the Kyoto Protocol: a Multi-Model Evaluation*. Special Issue of *The Energy Journal*.

Wigner, Eugene P., 1960. "The Unreasonable Effectiveness of Mathematics in the Natural Sciences," in *Communications in Pure and Applied Mathematics*, Vol. 13, No. I (February). New York: John Wiley & Sons, Inc.

Williams, Jeffrey, 1986. *The Economic Function of Futures Markets*. Cambridge: Cambridge University Press.

Williams, Jeffrey C., and Brian D. Wright, 1991. *Storage and Commodity Markets*. Cambridge: Cambridge University Press.

Williamson, Oliver, 1964. *The Economics of Discretionary Behavior: Managerial Objectives in a Theory of the Firm*. Englewood Cliffs, NJ: Prentice-Hall.

Wilson, R., 1985. *Introduction to Graph Theory*, 3rd edn. Harlow, UK: Longman Group Limited.

Wolfram, Stephen, 1999. *The Mathematica Book*, 4th edn. Cambridge: Wolfram Media/Cambridge University Press.

——, 2002. *A New Kind of Science*. Champaign, Ill.: Wolfram Media, Inc.

Yang, Z., R.S. Eckaus, A.D. Ellerman, and H.D. Jacoby, 1996. Report No. 6: "The MIT Emissions Prediction and Policy Analysis (EPPA) Model" (May), http://web.mit.edu/globalchange/www/rpt6.html#2.6

Index

lifetime of pleasing as Meins